COORDINATING VENTILATION

COORDINATING VENTILATION

SUPPORTING EXTINGUISHMENT AND SURVIVABILITY

NICHOLAS PAPA

Fire Engineering
BOOKS & VIDEOS

Disclaimer

The recommendations, advice, descriptions, and methods in this book are presented solely for educational purposes. The author and publisher assume no liability whatsoever for any loss or damage that results from the use of any of the material in this book. Use of the material in this book is solely at the risk of the user.

Copyright © 2021 by
Fire Engineering Books & Videos
110 S. Hartford Ave., Suite 200
Tulsa, Oklahoma 74120 USA

800.752.9764
+1.918.831.9421
info@fireengineeringbooks.com
www.FireEngineeringBooks.com

Senior Vice President: Eric Schlett
Operations Manager: Holly Fournier
Sales Manager: Joshua Neal
Managing Editor: Mark Haugh
Production Manager: Tony Quinn
Developmental Editor: Chris Barton
Cover Designer: Jared Hood
Book Designer: Robert Kern, TIPS Publishing Services, Carrboro, NC
Cover Photo: Brian Grogan

Library of Congress Cataloging-in-Publication Data

Names: Papa, Nicholas, 1987- author.
Title: Coordinating ventilation : supporting extinguishment and survivability / Nicholas Papa.
Description: Tulsa, Oklahoma, USA : Fire Engineering Books & Videos, [2021] | Includes bibliographical references and index. | Summary: "A training text for firefighters on techniques for ventilation"-- Provided by publisher.
Identifiers: LCCN 2021018787 | ISBN 9781593704377 (paperback)
Subjects: LCSH: Fire extinction--Technique. | Ventilation. | Air flow.
Classification: LCC TH9327 .P37 2021 | DDC 697.9/2--dc23
LC record available at https://lccn.loc.gov/2021018787

All rights reserved. No part of this book may be reproduced, stored in a retrieval system, or transcribed in any form or by any means, electronic or mechanical, including photocopying and recording, without the prior written permission of the publisher.

Printed in the United States of America

2 3 4 5 6   25 24 23 22 21

Dedicated to the life of Cade D. Townsend IV.
May this book and my actions honor you always.

CONTENTS

Foreword ix
Preface xi
Acknowledgments xiii

1 Training and Implications 1
   The Human Factor 3
   Learning the Hard Way 7

2 Fire Behavior 9
   The Fire Triangle 9
   Fuel and Heat 10
   Heat and Oxygen 11
   Oxygen and Combustion 12
   "Nothing Showing" 13

3 Ventilation Defined 21
   Air Flow and Fire Spread 22
   Inlets and Outlets 23
   Ventilation Limitations 27
   Ventilation Timing 28
   Tenability 30

4 Objectives and Methods 35
   Venting for Extinguishment 35
   Venting for Search 36
   Venting for Access 37
   Vertical Ventilation 38
   Horizontal Ventilation 43
   Hydraulic Ventilation 47
   Mechanical Ventilation 49
   Positive Reinforcement 50

5 Size-Up and Decision-Making 53
   The RADE Loop 53
   Fireground Tempo 58
   Learning the Hard Way 62

6 Operational Principles 67
Tactical Ventilation 67
Communication 68
Coordination 70
Control 72
Positive Reinforcement 74

7 Operational Practices 77
Guideline 1: Communication 77
Guideline 2: Door Control 80
Guideline 3: Wind and Access 84
Guideline 4: Wind and Ventilation 88
Guideline 5: Exposures 90
Guideline 6: Vent for Extinguishment—Horizontal 92
Guideline 7: Vent for Search—Horizontal 95
Guideline 8: Vertical Ventilation 103
Guideline 9: Void Spaces 111
Guideline 10: Positioning and Egress 116

8 Mission-Oriented Operations 119

Glossary 123
Notes 133
Additional Reading 145
Index 147
About the Author 155

FOREWORD

This book is a great reminder of how ideas can be brought to life, and it represents many years in the making. It's now a valuable contribution: a contemporary deposit to the fire service that benefits all of us who share in that labor. I am so delighted to see a book of this caliber, something unique and comprehensive, yet very firefighter-friendly and void of overcomplexity.

The author, Nicholas Papa, could not have tackled a more difficult subject. Ventilation has always been a subject of controversy and debate. Perhaps it goes against the grain or is contradictory to what we were taught, leaving us stuck somewhere between our traditional beliefs and the modern-day science of fire dynamics. This book demystifies the tactic of ventilation, perhaps debunking the myths and misconceptions in light of new empirical research and data on the impact of ventilation on the fireground. A straightforward approach of elaboration through simplicity truly sets this book apart from others: a cerebral delight designed to bolster your knowledge when it comes to ventilation.

This book represents a colossal dynamic interaction between knowledge, experience, and "the science," neatly arranged and packaged in such a manner to help us consistently make better, more intelligent decisions on the fireground. Timeless quotes and sayings from our mentors further reinforce and validate what we knew then, but what we know now really has not changed much, if at all. While the environment has certainly evolved, there has not been a radical paradigm shift in terms of how ventilation works; the physics remain the same. Today, we have more knowledge without assumptions and anecdotal theories. Nicholas Papa goes well beyond exceeding our expectations, which he has been doing for several years now.

As the fireground continues to evolve, so should we. Knowledge should be complementary to experience so that one does not overshadow or impede the other. The adage, "Try to learn something new each day," is timeless advice. Reading this book is a testament of that commitment. For me, with 25 years in the fire service, I am still a student, just as eager to learn as I was the first day I entered the Chicago Fire Academy. This book is rather appealing to the intellect for those wanting to know more, to learn. It's an investment that pays long-term dividends. Never stop learning!

I am honored and greatly humbled to call Nicholas my brother, friend, and mentor. I have a deep admiration for his dedication, diligence, and relentless pursuit to inform and educate our fellow brothers and sisters in the fire service. He is an embodiment of a true fire service educator in every sense. I offer him a sincere "thank you" for *tactically* assembling such a creative masterpiece for us all to enjoy. My brother, you have left one heck of an indelible impact upon me.

—Captain Jimmy Davis
Chicago Fire Department

PREFACE

I borrow the words of retired Commandant of the Marine Corps, General A. M. Gray, from the opening of the USMC organizational doctrine, *FMFM-1: Warfighting*, to describe this book: "By design, this is a small book and easy to read.... It is not intended as a reference manual, but is designed to be read from cover to cover.... The thoughts contained here represent not just guidance for action, but a way of thinking in general."[1]

Chapter 1 begins by explaining the training paradigm that exists in the US fire service and the effect it has on performance. To alleviate the associated discrepancies, chapters 2 and 3 provide a fundamental knowledge base of fire behavior and ventilation, identifying the relationship and the impact our tactics have on the fireground. Chapter 4 outlines the purpose and modalities of ventilation to demonstrate the context and its practical application. A decision-making model is then detailed in chapter 5, breaking down the components and variables and honing the size-up process to allow a more accurate determination of the appropriate course of action. Chapters 6 and 7 include a set of universally applicable principles and practices to serve as a guide for all ventilation operations, whether you work in the smallest rural community or the largest metropolitan city. The content and its purpose are solidified in chapter 8 by defining our core mission: the very basis of our actions.

The focus here is not on the specific techniques of ventilation—detailing the *how*—but is aimed at addressing the *when*, *where*, and more importantly, the *why*. Everyone's situation is inherently unique, from building stock to resources and everything in between. With such a wide range of dynamic factors, each incident must be addressed accordingly. While this book "provides the broad guidance" for doing so, it "requires judgment in application."[2] With a functional understanding of ventilation and a framework for executing operations, you can then consistently make the right call for *your* fireground.

The intent of this book is to pay it forward by passing along the insights and understanding I have developed to help enlighten and guide others as they operate on the fireground. By compiling more than 100 different references into a single, comprehensive document, this book allows the reader to not only garner knowledge, but more importantly, develop a *practical* understanding of ventilation. In sharing my experiences, from both the fireground and as a UL-FSRI technical panelist for the coordinated fire attack study, the material is given context and taken beyond the conceptual/theoretical realm. Through this, I hope to enhance your performance, embolden aggression, and prevent avoidable loss—upholding our oath and optimizing our service to the citizens and the communities we swore to protect. As legendary Chicago Captain Richard "Dick" Scheidt used to say, "Always leave the job a little better for the next guy."

—Lieutenant Nicholas Papa
Engine Company 1, New Britain (CT) Fire Department

ACKNOWLEDGMENTS

None of this could have been possible without the unconditional love and support from my wife, Regina, and my two kids, Giuliana and Thomas. Your patience, understanding, and sacrifice, despite my time away and countless hours grinding and toiling behind my laptop, allowed me to accomplish this lifelong goal. Thank you for sharing me with the fire service and always encouraging me to not just chase my dreams *but make sure I catch them*.

This work is the culmination of five years of extensive research and development, the experiences I have gained throughout my career, and a childhood and adolescence spent largely in the fire house and on the fireground riding along with my father, Deputy Fire Chief Frank Papa Jr. I have been incredibly fortunate with the opportunities that have been presented to me, and more importantly, with the many individuals who have influenced and supported me, both professionally and personally. There are simply too many of you to list here, but you are all embodied within these pages, and I will be forever grateful.

Thank you to the many talented and passionate photographers and firefighters who graciously shared their images, which brought these pages to life. Special thanks to Mark Haugh, Chris Barton, Tony Quinn, and Tiana Wendelburg from Fire Engineering Books & Videos; Steve Kerber and Craig Weinschenk from Underwriters Laboratories–Firefighter Safety Research Institute (UL-FSRI); Fire Chief Raul Ortiz and Lieutenant Shane Burns from the New Britain Fire Department; and Chicago Fire Captain Jimmy Davis, FDNY Battalion Chief Jerry Tracy, and Harwich (MA) Fire Chief Dave LeBlanc for all of their support with this publication.

1

TRAINING AND IMPLICATIONS

In tactics, the most important thing is not whether you go left or right, but why you go left or right. . . . We should base our decisions on awareness rather than on mechanical habit.

—General A. M. Gray,
USMC

The fireground is a dynamic and hostile environment, thrusting firefighters into time-compressed situations where life and property hang in the balance. These instances require the immediate selection and implementation of various tactics to properly intervene. This process takes place in mere seconds amidst varying degrees of chaos and stress, and it ultimately determines the fate of an incident. A *practical* knowledge of the mechanics behind each tactic will allow a more accurate prediction of the potential impact of the tactic on the fire, the environment, and the people within it, especially the unprotected victims. The ability to rapidly evaluate a situation and forecast the progression of the conditions is the very foundation of sound decision-making and successful execution. It is the possession of this mental skill set that distinguishes the laborers from the true craftsmen. As FDNY Lieutenant Joey DiBernardo would say, "Are you on The Job, or are you *into* The Job?" (fig. 1–1).

Throughout much of the fire service, there has been a common deficiency when it comes to the manner in which we educate our firefighters, especially new recruits. While we largely do an adequate job of training on the fundamental skill sets needed to physically perform the core duties, we often fail to correlate those tactics to fire behavior and the immediate effect of the tactics on the fireground, especially relating to victim survivability. The baseline introductory curriculum typically consists of a mere four hours of fire behavior lecture, usually scheduled early on in the program. The instruction is based on the standards of the

2 Coordinating Ventilation

Figure 1–1. Lt. Joey DiBernardo, FDNY
Courtesy: Matt Daly

certifying agency, preparing students for the examination rather than reality. More times than not, the material is not referenced again until the *live burns*, the culmination of training in which recruits are given a brief demonstration of the stratification of smoke/thermal layering and fire growth.

The live fire training most receive is unrealistic and does not adequately prepare the students for what they will face out in the field. These fires typically take place in a noncombustible structure, with little compartmentalization and no void spaces, and are strictly limited by the fuel source provided (fixed-sized stacks of pallets and hay). Not only does this inaccurately reflect the conditions within an actual structure fire, but the characteristics and output of the materials burning, including the building itself, are vastly dissimilar. The impact of the tactics performed, particularly ventilation, will result positively under these circumstances regardless of coordination, as the fire cannot get any bigger and the building cannot burn. The false sense of reality this provides can improperly code new firefighters as to how their tactics actually function and what they will encounter on the fireground. This can set them up for failure if the evolutions are not closely regulated and the students are not prompted by the instructors.

Being a highly physical vocation, the fire service has historically focused the greatest amount of its attention on the hands-on side of the trade. While continuously developing the capabilities to perform the essential tasks on the fireground is of the utmost importance, possessing a practical understanding of their function is critical to *consistently* achieving the best possible outcome. Ensuring the connection is made between the psychomotor and cognitive aspects requires that the fundamental principles of each tactic be instilled throughout all facets of training. In his book, *Warrior Mindset*, Dr. Michael Asken writes, "The mental and physical are two faces of the same coin and cannot be separated....We must realign our focus and esteem the mind every bit as much as we esteem the physical capabilities and skill sets."[1] Even when conducting the most basic drills related to a single skill or technique, firefighters must be provided with the context and the implications of conducting that operation in the field, reinforcing the cause-and-effect relationship it has with the fireground at large.

When firefighters comprehend the *why*, they can grasp the tactical options at their disposal and are then capable of making an informed judgement call. It takes them beyond the realm of simply knowing *how* and reacting purely out of habit or procedure. "The *why* is the missing component in the fire service....When we understand the why, we can solve the problems that are going to occur on the fireground."[2] Competency and performance are as much about the mental component (the "software") as they are about the physical (the "hardware").[3] By overlaying tactical experience with the laws of chemistry and physics, an optimal balance can be achieved, fusing the art of firefighting with the science of fire (and fluid) dynamics.

The Human Factor

With the Internet full of videos showing ventilation precipitating rapid fire growth and spread, along with other unfavorable outcomes, is the tactic to blame? Despite the litany of negative examples that are readily available and the subsequent fearmongering (particularly surrounding vertical ventilation), the answer is quite the contrary. Ventilation is a highly beneficial intervention that can even be the deciding factor for an incident. The linchpin for its success, however, is the individual executing the operation—the human factor. The fireground is a fast-paced and ever-changing environment, where the conditions and associated needs are variable and will influence ventilation and the manner in which it is addressed. Because of these complexities, firefighters must understand where they are fallible so they can make preparations and take the necessary precautions to navigate these pitfalls and weaknesses, minimizing their chance of failure.

Just like professional athletes, firefighters must watch "film" to better understand their opponent (fire) and learn from the experiences of others. The key to *productively* examining fireground footage, especially when the performance was less than ideal, is to place yourself in the position of those individuals, and more importantly, to consider their viewpoint. By analyzing the incident from their lenses, you can better determine what they were actually thinking in that moment and what they intended to accomplish. Once you understand their thought processes and objectives, the root cause and contributing factors of any deficiencies can be teased out, identifying mistakes and shortcomings so we can prevent them from occurring again. An examination of many cases of "ventilation gone wrong," including my own experiences, revealed four behaviors that repeatedly surfaced as the main and underlying causes: *overzealousness, freelancing, unawareness,* and *misguidedness*.

Overzealousness

Overzealousness is commonly associated with more junior or inexperienced firefighters. Whether it is one of their first jobs or the first time they are performing a particular function or task, some individuals may be overly anxious to get into the fray and execute. When smoke and/or fire is showing and they are in position with a tool in their hands, that "itchy trigger finger" may get the better of them and cause them to jump the gun. This lack of patience is caused by a loss of composure and is a sign of operational immaturity. Firefighters must understand that timing is the cornerstone of coordinating ventilation operations. Creating an opening prematurely or belatedly can be the catalyst for the overall failure of an incident, or at the very least, undue loss and injury.

Freelancing

Freelancing—any unsolicited and uncoordinated actions that do not adhere to organizational/best practices or promote mission accomplishment—should be prohibited and prevented. While the occurrence of this potentially detrimental behavior has seen a marked decline with the advent of the incident command system and the majority of agencies assigning every firefighter a portable radio, it is a problem that still very much exists. This lack of accountability may result from poor organizational structure or weak command presence. Through the establishment of operational systems, including standard operating procedures or guidelines, the *baseline* expectations and responsibilities can be laid out ahead of time. Predetermined unit and riding assignments that specify the roles of each company and member greatly facilitate continuity and integrity. This fireground "playbook" must be reinforced through training and must also be upheld at all incidents. Company-level commanders and incident commanders need to ensure the rationale behind these systems and their importance are understood. These behaviors must become the standard within an organization. The systems, however, should not be restrictive or promote micromanagement. There must be some flexibility in their construction to provide subordinates with a reasonable degree of discretion to make *justifiable* adjustments based on the conditions or circumstances encountered. Firefighters should be empowered to seize opportunities to better accomplish their tactical objective or to address a strategic priority, especially in matters of life safety. Lieutenant Gregory Turnell of the Washington DC Fire Department said it best: "The rules are never more important than the mission."

Unawareness

There are varying degrees of unawareness, ranging from not picking up on subtle details or events all the way to a breakdown of one's faculties, distorting perception of time and space. The most seemingly obvious cause of this behavior is a lack of vigilance. While inattentiveness or a lackadaisical mindset can certainly be contributing factors, the heart of the issue is often much deeper. Missing information or not realizing certain events are taking place can be largely the result of a training deficiency. Individuals who do not possess the essential background may not know how to properly assess the environment and interpret the situation. As a result, they may not understand or even recognize what is taking place. As the old saying goes, "You don't know what you don't know." The other major aspect of this problem ties back to composure. The more excited or anxious an individual becomes, the greater effect it will have on his or her situational awareness. Once someone's heart rate increases beyond the effective threshold, the individual will start to become impaired, which includes suppression of the senses (tunnel vision and auditory exclusion).[4]

Misguidedness

Misguidedness is likely the most prominent of the four main factors, as a practical understanding of how ventilation actually works is commonly lacking (fig. 1–2). Inaccurate assumptions may be made regarding the impact of ventilation based on our experiences and what we *perceive* is occurring on the fireground. In lieu of the facts, these anecdotal interpretations can become the standard within an organization. They can serve as the driving force behind the selection and implementation of ventilation tactics and have the ability to transcend generations of firefighters. Without a functional level of competency, an individual may overlook

6 Coordinating Ventilation

Figure 1–2. Firefighters aggressively open up the ceiling to expose concealed fire above for the engine company.
Courtesy: Brian Grogan

a critical aspect or variable and improperly execute an operation, despite the best of intentions. The impact of our tactics must be understood in order to take the proper precautions and ensure our interventions yield the best possible outcome for both the firefighting efforts and, more importantly, victim survivability.

For those very reasons, many agencies elect to task their more experienced firefighters with performing ventilation functions, assigning those firefighters to the outside vent and roof positions. Operating autonomously (independent of the company officer) in remote locations, often above the fire and without the protection of a handline, they are responsible for selecting the approach and the methods they feel will support the overall mission, given the circumstances before them. They must be able to perform an ongoing, macro-level assessment of the incident since ventilation is intertwined with all other aspects of the fireground. Ventilation will directly impact the interior operations and especially potential victims.

Successful assessment of the situation requires an in-depth understanding of fire behavior and building construction to locate the seat of the fire, determine its stage and extent of involvement, and identify the likely avenues of fire spread and its rate of progression. It also necessitates the discipline and mental acuity to accurately monitor the radio traffic and observe the changes in the conditions—gauging the operational tempo and tracking the progress of the incident.

Additionally, these individuals must be intimately familiar with the tactical options at their disposal. Understanding how each intervention works and anticipating what the resultant effects will be are key to making informed decisions. They can ensure the primary objectives are most effectively and efficiently supported.

Learning the Hard Way

Shortly after sliding the floor to the truck (after four years on the engine), I caught a good job in an occupied multiple dwelling. The rear porches of a six-family tenement (a three-story, ordinary-constructed building with two "railroad flat" apartments per floor) were fully involved and had extended to the interior on the upper floors. When we arrived, as the second-due truck, we were assigned to search the second floor (fig. 1–3).

I was riding in the irons position that night and paired up with the company officer. We set off to search the south-side apartment, while the other half of our crew searched the north side. The first-due companies were heavily engaged, combating heavy fire conditions on both the top floor and cockloft (after knocking down the porches). The other second-due engine was hooking up to the hydrant, and the third-due engine was just arriving on scene. Additional lines, therefore, had not yet been stretched. This left us temporarily operating *without any immediate protection*, a fact that should have played a major role in my future decision-making.

When we entered the apartment, visibility was limited to less than 1' off the floor, and a glow could be seen at the end of the hallway. We headed straight for it to start our search as close to the fire as we could and then work our way back. As we arrived at the far room, which ordinarily would have been a bedroom, we split up. With my captain anchoring at the doorway, I began searching. I noticed that there was no door, which should have raised another red flag. I could just

Figure 1–3. Street view
Courtesy: NBC CT

barely make out the shadowy outlines of the furniture, and it became clear the space was being used as a TV room. When I reached the far (outside) wall, I felt upward and discovered a window. I reverted back to my basic search training and remembered the words, "Vent as you go." I thought, "Great, I can take this window, the smoke will lift, and I'll be able to see better and move faster." As soon as I vented the window and got back down on my hands and knees to restart the search, I heard the captain calling, "Nicky, get back to the door, it's coming down the hall!"

As I made the next turn to quickly finish off the search on my way back, I could see the fire starting to roll across the ceiling just outside the doorway. The captain and I married up and quickly proceeded back down the hallway. Along the way, we fortunately crossed paths with the second-due engine company advancing in with a charged handline. As they drove back the fire, we worked through the living room and front bedroom to continue our search. By the time we finished, the engine had knocked it down, and shortly thereafter, the fire was brought under control (fig. 1–4).

On the whole, the incident was an overwhelming success, especially given the conditions on arrival. It was a great stop, in large part due to the competency and aggressiveness of the members of the first-due companies. Lost in the victory, however, was the lesson I should have learned immediately after venting that window. It was not until several years later, while attending a fire behavior lecture, that I would finally have that epiphany moment and realize the implications of my actions—I had caused that event. This incident reinforced the fact that "experience and knowledge have to coexist; you don't gain experience if you don't know what is happening."[5]

Figure 1–4. The rear porch where the fire started, extending to the interior
Courtesy: NBC CT

FIRE BEHAVIOR

Constantly consider that ventilation will cause the fire to grow rapidly and will increase the temperature in the compartment. Even ventilation above the fire will accelerate the burning rate by allowing fresh air to replace the oxygen deficient atmosphere that was slowing combustion.
—Battalion Chief William Clark, FDNY

The Fire Triangle

Fire is defined as the rapid oxidation of a material through the process of combustion. When heat is applied to a combustible fuel, the material begins to break down in a process known as *pyrolysis*. The thermal decomposition of the fuel causes it to off-gas and vaporize, which begins at less than 392°F (200°C).[1] On reaching its respective ignition temperature and supplied with the proper concentration of oxygen, flaming combustion can be achieved. The fire triangle represents the mutual relationship between the three fundamental parts of the combustion process (fig. 2–1).

The three components—fuel, heat, and oxygen—collectively form a chemical chain reaction, prompting the creation of an additional term, the *fire tetrahedron*. Each of the components is dependent on the others. If even a single leg is compromised (including the reaction itself), the combustion process will be inhibited. While oxygen is the component most closely linked with ventilation, all of the components are interrelated and therefore equally important to understand.

Figure 2-1. Fire triangle

The basic premise of this most fundamental fire behavior concept is often lost in the training and education of firefighters and is not practically translated to the fireground. As a result, the initial actions of firefighters can actually result in conditions worsening, at least initially, particularly in regard to ventilation. When executed tactically, ventilation can greatly enhance all other fireground operations and victim survivability. It can even be the deciding factor for a positive outcome. Despite the potential value and longstanding status as a foundational firefighting tactic, it is, unfortunately, "one of the most *poorly* practiced."[2] To ensure ventilation and other operations are successfully executed, an understanding of how the elements of the fire triangle/tetrahedron are connected and how they impact the fireground will be paramount.

Fuel and Heat

As fuels are consumed in the combustion process, energy is produced in the form of heat. Every fuel has a defined heat capacity, known as its *potential energy*, which is determined by its chemical makeup.[3] It is widely known that synthetic fuels (plastics) have a much greater energy potential than their natural counterparts (wood and organic fibers), producing in excess of twice the amount (approximately 8,000 versus 16,000 British thermal units [Btu] per pound). Although this is certainly important, it is the increased speed at which the synthetics liberate their heat energy that is of far more concern. Synthetics release heat at a rate of up to four times greater than that of natural materials.

Not only do these fuels increase the fire load and cumulative heat output, their drastically higher heat release rate (HRR) accelerates the combustion process and the deterioration of the conditions within the involved area, promoting rapid fire development and reducing tenability when provided with a sufficient supply of oxygen.[4] For this reason, there can be an appreciable difference between a fire in a living space involving mostly synthetic contents and a fire within the attic/cockloft and knee-wall spaces involving mostly natural structural members. The considerable disparity between the two fuel loads can result in a significant difference in the confinement ability of ventilation and its associated window of effectiveness prior to water application. The increased speed and voracity with which synthetic fuel loads react to additional sources of oxygen require greater diligence in the timing of *all* forms of ventilation within living spaces. It must also be noted, however, that the by-products of those synthetic fuels can migrate through the void spaces of a building and dangerously accumulate in the concealed spaces above or adjacent to the occupied areas. In this case, prompt vertical ventilation will exhaust the hot fire gases and smoke, relieving the mounting pressure and the potential effects of a *cockloft (smoke) explosion*. Synthetic fuels may also be stored in the attic/cockloft or knee-wall spaces, particularly in larger, walkable areas, which requires that the timing of ventilation be well-coordinated and disciplined, regardless of where it is being conducted.

Heat and Oxygen

For fuels to achieve their peak heat release rates and liberate their full energy potential, a sufficient oxygen supply must be available to the fire. In 1830 James Braidwood, the first superintendent of the London Fire Brigade (LFB), identified this connection. He noted that "the door should be kept shut while the water is being brought, and the air excluded as much as possible, as the fire burns exactly in proportion to the quantity of air which it receives."[5] Nearly a century later in 1917, British scientist W. M. Thornton determined that for every unit of oxygen consumed for combustion, there is a fairly constant heat output among common (organic) materials. This relationship between oxygen and heat output, known as *Thornton's rule*, states that 13.1 megajoules (MJ) of energy are released per 1 kilogram (kg) of oxygen.[6]

In 1980, NIST researcher Clayton Huggett conclusively validated and refined this work using calorimetry. By modifying the equation to include the percentage of oxygen found in ambient air (21%), he was able to obtain a practical baseline measurement of the oxygen/heat relationship: 2.75 megajoules (MJ) of energy per 1 kg of air, or 1,180 Btu of energy per 1 pound (lb) of air (US customary units).[7]

In more practical terms, 1 cubic foot (ft^3) of air (weighing 0.08 lbs) will yield 95 Btu of heat. As a point of reference, a typical bed produces 950 Btu/sec (1 megawatt [MW]) of heat. At that rate, the fire would consume 10 ft^3 of air per second. In a 12' × 12' × 8' room (1,152 ft^3)—a common size for a residential compartment—the available oxygen within the baseline volume of air would be consumed in just under 2 minutes (115 sec) *for that fuel package alone.*[8]

It is important to note, however, that the supply of air within a room is not fixed, as compartments within a building are interconnected by doorways and entryways, and windows to the outside atmosphere may be open or may fail. The oxygen concentration is also highly variable throughout the course of a fire. As the fire grows, so will its rate of oxygen consumption, increasing the speed with which the supply can be rendered insufficient to support combustion. These findings have brought attention to the significance of the oxygen supply, as well as the effects resulting from how the fire environment has evolved. Oxygen supply will impact not only fire growth and spread, but the survivability of victims as well.

Oxygen and Combustion

The supply of oxygen available for combustion is determined by a number of environmental factors, including the size of the compartment, the available openings and their orientation, the wind velocity, and the subsequent flow of air.[9] When a fire grows and its output increases, so does its demand for oxygen. As the concentration of oxygen within the space begins to diminish, the fuel load's ability to release its potential energy will be equally reduced. Under typical residential fireground conditions, if the oxygen concentration falls below 15%, the fire can stop flaming combustion altogether and begin to smolder.[10] Based on the average heat release rate experienced from a common contents fire, the oxygen supply in an enclosed residential compartment can be consumed in just a few minutes.[11] Even in the commercial fireground setting, where the volume of the space can be many times larger (as well as the fuel load), the rate of oxygen consumption can quickly deplete the available supply, becoming underventilated and limiting the heat release rate. This may occur within as little as five minutes.

The combustion process is not linear, and the volume of air is variable since additional supply may be provided from external openings and adjoining compartments. Even so, these examples serve to demonstrate how rapidly the fuel load can consume oxygen. In the absence of the ideal air flow, the fire will remain underdeveloped and is known as a *ventilation-limited fire* (fig. 2–2). As oxygen within the involved compartment drops, the fire will begin to draw air in from adjacent open compartments, producing a negative pressure in the lower portion of those spaces. The upper portion, however, results in a positive (higher) pressure, as it is filled with hot combustion gases. The greater pressure above exerts downward force on the fresh air below, which further facilitates its migration toward the fire compartment. This is predominately the case with today's fires, given the fuel-rich and energy-efficient

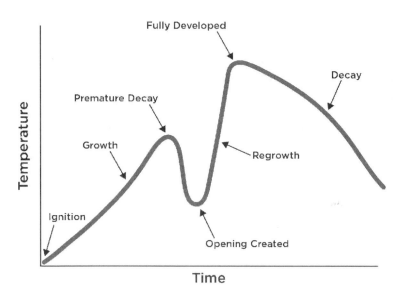

Figure 2–2. Growth curve of an underventilated or "ventilation-limited" fire

environment, increasing the oxygen demands while restricting its supply. The majority of growth phase fires are underventilated when the fire department arrives and should be treated as such.[12] When the air supply becomes insufficient, the fire will enter into a premature/ventilation-induced decay stage, suppressing its development and output and eventually self-extinguishing between 13% to 15% oxygen concentration.[13]

A fire acts similarly to a human when suffering from oxygen deficiency.[14] At first, it will begin to "hyperventilate," behaving erratically while it attempts to compensate for the lack of oxygen, resulting in the fire producing a high-velocity, pulsating exhaust. As it runs out of energy, the exchange slows, and the fire "gasps" for air—attempting to intake through the outlets—before eventually stopping altogether, resulting in extinction. An underventilated fire is, in essence, "hypoxic" combustion. With the fuel load and the heat still present, however, this condition can pose a threat if the necessary precautions are not taken, setting the stage for rapid fire development and potentially resulting in a ventilation-induced flashover or even a backdraft.[15]

"Nothing Showing"

In the time it takes for the fire to be detected, notification to be made to 9-1-1, and the fire department to be dispatched and to respond, underventilated conditions can be encountered. When the fire department arrives on scene, the fire may have decayed to the point that smoke is no longer exhausting from the building. The lack of oxygen causes the combustion process to become unstable and eventually leads to self-extinguishment. As the energy (heat) output diminishes, the resultant decrease in temperature causes the gases to contract. As the pressure subsequently drops, possibly even below ambient levels, it will become insufficient to discharge the heat from the available outlets.[16] This may prompt the common on-scene report of "nothing showing." That phrase should not alter the focus or preparedness of the crews operating or responding. It simply means that there is no immediate indication of fire from that initial position (fig. 2–3).

Further investigation is needed, including a complete exterior and interior survey of the building (commonly referred to as a *360-degree size-up*). What may be present in the rear may not be readily visible from the original vantage point. The first-arriving crews must do their due diligence, enlisting all of their senses to search for other clues or indicators of fire. Tell-tale signs of fire could include blackened or crazed/cracked glass, condensation buildup on the inside of the glass (informally known as "weeping windows"), soot-staining or charring around openings, and the odor of burning[17] (fig. 2–4).

Firefighters must resist the seduction of complacency and maintain their situational awareness, communicate any pertinent findings, and exercise the proper precautions. Because today's fires are so overly fuel rich, they will inherently deplete the oxygen supply well before the fuel load is ever consumed. With that in mind, these incidents where "nothing is showing" should be approached as though an underventilated fire condition exists in the building. Firefighters must acknowledge and appreciate the power of ventilation and operate in a state of readiness and vigilance throughout the duration of the incident.[18] Taking your eye off the ball for even a second could prove detrimental or even deadly.

Smoke

Smoke is one of the primary targets for ventilation efforts, so it is critical to understand the nature of smoke and its associated impact on the fireground. Smoke is the by-product of *incomplete* combustion: a collective of solid particulates (carbon, ash/soot), aerosolized liquids (hydrocarbons, oils/tar, water) and gases (carbon monoxide, hydrogen cyanide, benzene, acrolein). These by-products become elevated in a column due to the heat increasing their buoyancy, an effect known as a *thermal updraft*.[19] Because it is comprised of unburned fuels, smoke is readily ignitable when its parts are within their respective ignition temperatures and concentrations. While the fire hazard is certainly a major threat, the toxicity of smoke is the chief concern when it comes to victim survivability. It has been crudely said that "you are smoked before you are fried." The aforementioned gases

Figure 2–3. Smoke fills the opening as it exhausts from the second-floor window.
Courtesy: NIST

Figure 2–4. The fire has decayed to the point that it lacks the energy to exhaust the smoke.
Courtesy: NIST

are potentially fatal, and lethal doses can be readily achieved, at least to some degree, at almost every building fire. The less efficient the burning (flaming combustion), the more smoke will be produced. Given the high fuel-loading (predominantly comprised of low-mass synthetics) and energy-efficiency of today's environment, fires are progressing and depleting the oxygen supply more rapidly, becoming underventilated and notably increasing the smoke output.[20]

When smoke in the proper air/fuel mixture is heated above its flashpoint, it can begin to ignite. This is observed at the ceiling level (the *hot gas layer*) and initially presents as flickers or fingers of flame known as a *rollover*, an imminent precursor of a flashover. While this is a consistent indicator, it may not be readily observed from below, as the magnitude of the smoke may completely obscure it. Firefighters must enlist their other senses and utilize thermal imaging cameras to properly monitor the conditions. In some cases, the smoke possesses the necessary heat to ignite, but the amount of oxygen present is insufficient. As openings are made and the smoke migrates to the areas of low pressures, however, it can lean out as it disperses. The smoke can then ignite as it mixes with fresh air and enters the flammable range. This may not occur until the smoke discharges from an exterior opening into the outside atmosphere, in which case it is known as *vent-point ignition*.

Smoke is fundamentally evaluated on the basis of four physical characteristics, identified by Battalion Chief David Dodson in his book, *Art of Reading Smoke*: volume, velocity, density, and color.[21] The presentation of smoke can be a tremendously beneficial gauge of not only the present conditions, but also how they will change. Accurately performing this assessment requires an understanding of the definitions for each of the aforementioned attributes and how they respectively correspond to fire behavior. Effectively executing the process necessitates a vigilant mindset and a keen eye, particularly on arrival, resisting a narrow, task-driven focus (the "moth-to-flame" effect) and maintaining a much broader perspective. While visible fire naturally captures our attention, a brief assessment is all that is needed initially to verify its intensity and the location of that area of involvement. What requires far more attention, however, is scanning the remainder of the building for smoke, examining its presentation, and attempting to identify the seat of the fire, its stage of development and extent, and the rate of change (fig. 2–5).

Volume

The first and most visibly obvious of the four smoke characteristics is volume. The amount of smoke present indicates the extent to which the fuels within the structure are off-gassing as a result of pyrolysis.[22] The production of smoke is relative to the type of fuels burning and the efficiency of the combustion process. Synthetic (hydrocarbon-based) materials will yield much more smoke than those naturally derived. In either case, however, the oxygen and heat greatly affect the respective volume of smoke being generated. Because smoke is the result of incomplete combustion, greater amounts will be produced as the oxygen concentration begins to drop. As the heat energy decreases, the fire will burn less efficiently and become underventilated. The size of the involved compartments and building as a whole must be taken into consideration when evaluating the volume of smoke.

Figure 2–5. The fire is within the void/attic spaces, involving the wooden structural members.
Courtesy: Jean-Michel (J. M.) Reed

The larger the space, the greater the significance since the amount of smoke required to fill it will increase comparatively.

Velocity

The velocity of smoke is the most telling of the attributes in regard to the severity of the fire. Smoke is discharged by the force exerted from the pressure built up inside, generated either by the gases expanding from the heat or by the volume of smoke occupying the available space. Differentiating between the two causes is accomplished by observing how the smoke behaves as it leaves the building. Heat-pressurized smoke will naturally rise and progressively lose its speed as it cools, while volume-pressurized smoke immediately begins to dissipate and mix with the outside air.[23] The velocity of the smoke may also be impacted by the wind. If smoke is exhausting straight out or at a downward angle, it is an indicator that the fire is being driven by the wind. The smoke may also be discharging sporadically if the wind is gusting.[24] When assessing the presentation throughout the building, keep in mind that the farther away the exhaust point is from the fire and the larger the opening, the slower the smoke will discharge. To accurately compare the smoke velocity at different points of the building, therefore, openings of similar dimensions should be utilized.

The final and most critical aspect of velocity is the uniformity of the smoke column, an indicator of the cumulative heat condition. If the smoke is discharging in a tight, streamlined fashion, known as *laminar smoke*, the building and its contents are still capable of absorbing heat. When the smoke is exhausting erratically, chugging and twirling (known as *turbulent smoke*), however, the compartment has become saturated to capacity. The excessive buildup of heat, projecting back onto the smoke, or *radiation feedback*, causes the gases within the smoke to rapidly expand; with nowhere to go, this is a sign that flashover is likely imminent.[25]

Density

The density of the smoke, commonly referred to as the *thickness*, displays its potency and fire spread potential (fig. 2–6). As with volume, smoke density indicates the quality of the combustion process.[26] The more inefficient the fire is burning, typically caused by a lack of oxygen given today's fuel-rich environment, the more unburned products will be present in the smoke, increasing the likelihood and magnitude of it igniting, as well as its toxicity. The higher the density of the smoke, the less tenable the atmosphere will become for potential victims, and the greater the chance there is for rapid fire progression.

Color

Evaluating the color of smoke is most effective for determining the stage of the fire.

Regardless of the fuel type, most materials will produce a white smoke when they are initially heated, due to the water or chemical contents vaporizing first in naturals and synthetics, respectively.[27] As the moisture is burned off and the material is further consumed, the color of the smoke begins to darken. Organic fibers can yield a brown shade that is typical of more advanced fires in the void

Figure 2–6. Thick, black smoke pours out of the windows and immediately rises up in a tight column.
Courtesy: Matt Daly

spaces, involving the wooden structural elements. Plastics and other hydrocarbon-based materials give off smoke with a gray hue that is common to growth-stage contents fires. For all fuels, however, the color will continue to deepen as it is heated until the smoke eventually becomes a high-density black due to the buildup of carbon particulates, known as *carbonization*. Furthermore, as the fire intensifies, directly impinging on other nearby surfaces, those fuels will immediately produce a low-density black smoke as the heat of the flames increases the soot content.[28] If the compartment is well sealed, however, the smoke can become a yellow/mustard color as the fire decays and begins to smolder, a warning sign of potential backdraft. In essence, the darker the smoke, the greater the temperature, with the hottest smoke being closest to the fire.

When evaluating the color of the smoke, it is also critical to consider the building size and the weather (fig. 2–7). The farther the smoke has to travel and pass through restricted openings, the more it will mix with fresh air and moisture and come into contact with other fuels and porous surfaces. As heat is absorbed, vapor and particulate components that provide the dark color are shed off in the process, and the smoke can become lighter.[29] If the fire is deep seated, especially given larger structures and/or heavy contents loading, the smoke can present atypically in this manner. Because the temperature of the smoke largely determines its color, cold weather can also alter its presentation. In subfreezing conditions, this "cold smoke" effect can cause it to appear significantly lighter.

Be sure to account for these variables, in addition to collectively factoring in the other three smoke characteristics, to accurately determine the present conditions and forecast their progression.

Figure 2–7. Cold weather and the smoke's travel path can alter its presentation.
Courtesy: Jean-Michel (J. M.)Reed

VENTILATION DEFINED

> *We are responsible for venting the fire building to support search and/or extinguishment efforts....It is not simply a cut-and-bust operation. We are professionals and select primary and secondary channels to improve conditions and positively impact the fire behavior, based on an ongoing size-up.*
>
> —Fire Chief Tom Brennan,
> Waterbury (CT) Fire Department

Ventilation is a vital fireground task intended to systematically remove the by-products of combustion and extinguishment from a structure and replace them with fresh air through the creation and management of new and existing openings. When executed properly, it supports the two primary fireground objectives of fire attack and search, as well as victim survivability. As the fireground challenges continue to evolve due to the abundance of synthetic materials and energy-efficient alterations, a number of detrimental effects are being experienced:

- High fuel loading and heat outputs
- Rapid fire growth and spread
- Decreased time to flashover
- Increased oxygen consumption and toxic gas production
- Underventilated fire conditions
- Greater volatility

Collectively, these conditions drastically reduce the fire's reaction time and the margin for error when conducting ventilation operations, requiring a more disciplined approach.

Air Flow and Fire Spread

If an opening is created without the prompt application of water or management of the resultant air flow, the intake of fresh air (rushing in at about 5 mph or faster) will cause the fire to reenergize in a second growth stage. Once the fresh air arrives, the exchange generally takes about 10 seconds.[1] Conditions can then start to deteriorate and eventually become untenable if left unchecked, which may occur within an average of 1.5 to 3.5 minutes in typical single-story and two-story private dwellings, respectively.[2] The speed and intensity with which the fire reacts to ventilation will vary depending on several factors:

- The conditions within the compartment (temperature, stage of the fire, layout, and fuel load/configuration)
- The characteristics of the openings (location and size)
- The weather (wind velocity)
- The resultant air exchange (pathway and type of flow)

The higher the interior temperatures, the greater the fuel load, the closer in proximity the openings are to the fire, the larger the opening(s), and the more efficient the air flow, the faster the fire will respond to the ventilation, which is amplified when impacted by a prevailing wind condition.[3]

Fire produces energy in the form of heat in what is known as an *exothermic reaction*. On being heated, the gases expand and generate pressure within the compartment. Conversely, gases contract on being cooled, proportionately decreasing pressure according to *Gay-Lussac's law*.[4] Gases migrate along a *pressure gradient* from areas of high pressure to those of lower pressure. Taking the path of least resistance, the gases travel along a *flow path* within the volume of space between the available openings created by entryways, window openings, stairways, and roof structures.[5] As the fire grows, increasing its demand for oxygen, the rising column of heated products of combustion (the *fire plume*) can entrain cooler air through any open doorway within the compartment, creating a negative pressure and drawing from the adjoining portions of the structure.[6]

It is important to note that ventilation is constantly occurring to some degree in all buildings. The air within the interior naturally exchanges with the outside environment air through leakage at structural connection points and at the seams around openings, as well as through heating, ventilation, and air conditioning (HVAC) systems.[7] In the mid-1800s, London Fire Brigade (LFB) Superintendent James Braidwood first brought this fact to light, originally identifying the pattern of air travel as the "draught" (draft).[8] As the fire develops, the heat and smoke will rise and expand outward, moving away from the area of high pressure and banking down near the walls of the compartment in an effect known as *mushrooming*. When an opening is created on the same level or above the area of involvement, the heat and smoke (along with the fire) will be drawn to that area of low pressure.[9] In 1972, FDNY Deputy Chief Emanuel Fried similarly noted, "Ventilation creates a draft effect and may pull fire either vertically or horizontally."[10] While the terminology may have changed over time, the fundamental fire behavior and ventilation concepts have been understood and documented in our industry texts for some time.

The fire service, like any other profession, can be overwhelmed by novelty under the guise of innovation. While often well intentioned, attempts at advancing or refining a particular aspect can result in merely a reproduction or rebranding of an existing concept with no apparent addition in value. These alterations may convolute or complicate a concept that was formerly simple or straightforward. This is commonly encountered when it comes to terminology or the creation of new phrases or "buzzwords," along with changes in their definitions. This semantical confusion is highly problematic when it comes to educating new firefighters. While our industry is greatly regionalized and jurisdictionally driven (as it should be), we must strive to develop and adhere to a standard language when it comes to the topics that are universal, such as fire dynamics and victim survivability. A common set of trade-specific terms and definitions (i.e., jargon) must exist in the fire service to allow its members to universally embody professionalism in their vocation.[11]

Inlets and Outlets

If an opening is above the fire's point of *thermal balance*—the horizontal dividing line that separates the heat and smoke rising/accumulating along the ceiling and the cooler air remaining along the floor (known as the *neutral plane* when observed through an opening)—it will serve as an outlet.[12] As the heat and smoke are exhausted, an inverse air flow reaction is produced. Any opening that is below the thermal balance will proportionately pull fresh air into the compartment, serving as an inlet. If an opening is located on the same floor as the seat of the fire (also known as *on level*) and encompasses an area above and below the thermal balance, it will typically function as both an inlet and an outlet, resulting in an opposing air exchange known as a *bidirectional flow*.[13] This is not a recent discovery. As Parkersburg (WV) Fire Chief Lloyd Layman noted in 1955, "Open or burned out windows can serve as both exhaust and air-intake openings. Heated smoke is exhausted from the upper section while cool air from the outside atmosphere enters through the lower section. This action also occurs through an open doorway."[14] Because the opening is performing the two functions simultaneously—the intake and exhaust competing for the same area—the rate of exchange is limited, reducing its efficiency (fig. 3–1).

When an opening exists entirely above the thermal balance, it will typically serve as a dedicated outlet, producing a direct air exchange (*unidirectional flow*), maximizing its exhaust efficiency (fig. 3–2). The higher the opening is positioned

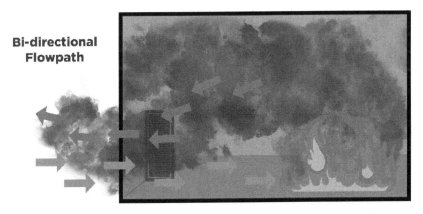

Figure 3–1. The door is on level with the fire and both above and below the fire's thermal balance.
Courtesy: FDNY

Figure 3–2. The exhaust opening is above level with the fire and the thermal balance.
Courtesy: FDNY

above the fire—the area where the highest temperatures are located—the more rapid the air flow/exchange will be, reducing the distance the by-products have to travel and capitalizing on the buoyancy and the pressure of the hot gases (the *thermal updraft*). Conversely, an opening situated fully below the thermal balance will function entirely as an inlet, as mentioned previously.[15] There is more to ventilation than merely the exhaust component. To achieve the desired effect, the intake must be accounted for as well. The inlet and outlet are mutually dependent on each other to sustain a continuous flow of air. When an opening is created, there is an "equal and opposite reaction" acting in accordance with Newton's third law of motion. As the heat and smoke are being exhausted, fresh air is proportionately drawn in.

The introduction of the additional air supply, the majority of which is typically entrained (remotely) through the open entry door, can initially improve visibility and victim survivability along the intake pathway at the floor level as it migrates towards the fire. This fact has been documented in fire service texts dating as far back as the mid-1800s. As LFB Superintendent James Braidwood noted, "When approaching a fire, it should be done by the door, if possible. When this is attended to, the current of fresh air, entering by the door and proceeding along the passages, makes respiration easier and safer than elsewhere."[16] Since the advent of self-contained breathing apparatus (SCBA) and encapsulating turnout gear, its importance has been largely forgotten or undervalued, as firefighters no longer rely on the relief obtained from the fresh air coming in through the entry door behind them. The magnitude and duration of that positive effect, however, will be conditional and highly variable based on the fire, the environment, and the subsequent firefighting actions taken.[17]

> Ventilation is not merely creating openings; it is equally about *managing* openings.

Because the vast majority of fires are underventilated, any additional supply of oxygen received will cause the fire to intensify until the onset of fire attack. While it may initially lift and redirect the heat and smoke above or opposite the advancing crews, improving conditions at the floor level, the openings created will not have the capacity to *sustainably* contend with the energy potential of the growing fire. Once the outlet is overwhelmed, any benefits gained along the intake pathway will begin to reverse as the fire conditions deteriorate. Consequently, it is imperative to limit the flow of air to the fire until the engine company is at the very least moving in and able to advance to the seat of the fire. Ventilation is not merely creating openings; it is equally about *managing* openings.

The flow of air and the by-products of combustion can also be altered by the fire suppression actions. When the handline is operating, the stream entrains air from behind it, through the entry door. The volume of air is increased when the straight or solid stream is agitated, most prominently in the "O pattern," producing roughly 5,000 cubic feet per minute (cfm or ft^3) at 150 gallons per minute flow. This surge of air being driven ahead creates a pressure front that is capable of occluding the approach corridor (i.e., the hallway), preventing the smoke and hot gases from exhausting overhead, tracking to the open door. The bidirectional flow previously occurring at that entryway can be converted into a unidirectional flow toward the seat of the fire. If the fire compartment is vented ahead of the handline's advance, this flow reversal can extend throughout the intake pathway and fully redirect the exhaust out from the opposing vent opening. Flowing and moving uses the handline's stream to leverage the air entrainment, in addition to surface cooling and gas contraction. It limits the spread of fire and its by-products, taking control of the space being traversed on the approach and reducing the thermal and toxic exposure for the firefighters, and more importantly, the unprotected victims. It is important to note, however, these benefits can be lost the moment the nozzle is shut down prior to extinguishment, potentially causing the conditions to immediately rebound.[18]

When observing the smoke exhausting from an opening, be sure to take note of the space it occupies. If there is a defined smoke layer allowing fresh air to intake below in a bidirectional flow, the position of the neutral plane can be helpful in gauging the stage of the fire. The lower the height of the smoke layer is within the opening (parallel with the thermal balance), the more severe the fire inside will be. As the conditions worsen, the neutral plane can become turbulent just prior to flashover, bouncing in an almost wave-like motion (fig. 3–3).

The neutral plane can also aid in determining the fire location and extent. When a defined neutral plane is present, the opening is likely on level with the fire. If the smoke is exhausting from floor to ceiling in a unidirectional flow—serving as a dedicated outlet—the opening is typically above the fire floor (also known as *above level*). This type of absent neutral plane, however, can also occur on level when a wind-impacted condition exists. If an opening is present on the upwind side, acting as an inlet, an opening on the downwind side can produce a pure exhaust. The same effect is also possible if the on-level opening, which is typically the entry door, is the only substantial point of ventilation and the fire has an abundant fuel load and an excess air supply. This can be the case in large, uncompartmentalized spaces such as commercial occupancies and even massive, open-floor-plan private dwellings, also known as "McMansions." Because of the potential fuel and oxygen availability, greater heat release rates can be experienced. The subsequent increase in temperature and gas expansion can generate enough pressure to overwhelm the limited opening. A unidirectional exhaust flow can be produced until the fire becomes ventilation-limited, requiring the opening to start intaking fresh air, transitioning to a bidirectional flow.[19]

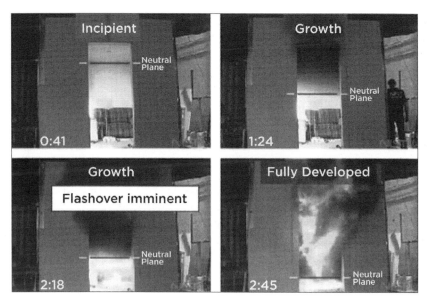

Figure 3–3. Neutral plane and stage of development
Courtesy: NIST

While creating additional ventilation points may briefly relieve the pressure, the increased intake of fresh air to the fire—leaning out of the environment and further increasing the heat release rate—will cause the fire to accelerate and spread toward the new low-pressure opening if it is not accompanied by suppression. As noted in the UL-FSRI coordinated fire attack study concerning strip mall fires, "Consideration should be given to the time it will take to apply water before any opening in the compartment is made."[20] For this reason, fires involving commercial occupancies require a much higher level of discipline, especially when it comes to the matter of horizontal ventilation, given the size and air flow capacity of the front display windows. These fires are far less forgiving, as the fire behavior and structural stability can deteriorate rapidly, particularly in type II noncombustible buildings, which are the most susceptible to collapse. Fires in these buildings demand their own specific approach. Attempting to attack a fire in a commercial occupancy with residential tactics can result in catastrophe.

Ventilation Limitations

Every opening has a finite exhaust capacity and therefore a limited ability to confine the fire and its by-products. As the fire develops, the energy it produces can quickly exceed the exhaust capabilities of the opening created, particularly with fuel-rich contents fires. After the opening has been consumed, the fire will start to spread laterally, seeking other low-pressure avenues, potentially at an overwhelming velocity. In 1974, FDNY Fire Chief Edward McAniff noted, "Once ventilation is started, the time for effective extinguishment is limited."[21] Given the fuel load and configuration of a typical dwelling compartment, there are not enough existing openings (approximately 70 ft^2 within an average living room) to realistically contend with the potential output of today's fuel load without being coordinated with extinguishment.[22] Fire Protection Engineer Hossein Davoodi determined that for ventilation to exhaust the total potential energy output of a fully developed compartment fire, roughly 25% of the wall area would have to be opened. Given the average residential fuel loading (0.07 MW/ft^2) within a common 12' × 12' × 8' bedroom, producing an estimated peak release rate of approximately 10 MW (9,500 Btu/sec), the required horizontal openings would total about 100 ft^2.[23] These figures are aligned with the UL-FSRI coordinated fire attack study data, which found that horizontal ventilation comprising an area of 10 ft^2, with the opening(s) on level and producing a bidirectional flow, has a heat release rate capacity of 1 MW (950 Btu/sec).[24]

Similarly, utilizing a standard residential hole size of 4' × 4' as a vertical ventilation opening yields a more efficient unidirectional flow. It is only capable of exhausting the heat release rate equivalent of a sofa fire, however, which is approximately 3 MW (2,850 Btu/sec), a mere fraction of the total potential energy of the fuel load.[25] For it to completely confine the fire under the same conditions, assuming the fire achieves optimum combustion and is fully regulated by the fuel load (known as a *fuel-limited fire*), an 8' × 10' hole would have to be made. This fact renders both approaches impractical choices to sustainably confine the fire without the prompt application of water, especially for contents fires. For this very reason, it is of the utmost importance that ventilation be coordinated with door control and the onset of suppression, managing the air flow and heat production.

Ventilation Timing

Preemptive ventilation may enhance conditions *momentarily* along the intake pathway by releasing some of the pent-up heat and smoke and drawing fresh, cool air into the compartment. This can promote a short-term lift of the smoke and heat by the inrush of air toward the fire, replacing the exhausting by-products in an effect known as *smoke tunneling*, temporarily improving conditions at the floor level, with the benefits increasing with proximity to the inlet.[26] If water is not promptly applied to the fire or if the flow of air is not managed until that time, however, the benefits will be short-lived (fig. 3–4).

Figure 3–4. The outside vent firefighter waits to take the glass.
Courtesy: Matt Daly

Premature ventilation can cause the fire to intensify and spread, allowing fuel-rich, heated gases to lean out and reach their ignition point, burning more efficiently. The "cleaner" burn produces less smoke particulate and additional light from the increase in flaming combustion. The resulting improvement in visibility that can occur may be misinterpreted as an exhaust-induced lifting of the thermal balance.

The outcome experienced will depend on the size and configuration of the building, the fuel load, and the openings, as well as the stage of the fire. If ventilation is executed too early, with the reflex time of suppression exceeding the duration of the air exchange and the reaction time of the fire (the "grace period"), the fire can grow to the point where it overwhelms the opening created and begins to deteriorate conditions. When the fire is left unchecked, the point of diminishing return for any ventilation tactic can occur rapidly. As the fireground evolves, the window of effectiveness for ventilation continues to narrow, reducing the margin for error.

The experiments for the UL-FSRI coordinated fire attack study, which were conducted in acquired structures with actual furnishings/contents to most accurately reflect a real fireground, demonstrated just how delicate a balance this dynamic truly is. Where preemptive ventilation actions were taken, occurring just 30–60 seconds prior to water application (a fairly conservative time frame), the fire was still able to grow and increase its output after the arrival of the fresh air within the involved compartment. What is far more pressing, however, is that "no meaningful increase in temperature *outside* the fire room was observed when ventilation tactics were executed in coordination with (shortly after or shortly before) the onset of suppression" (emphasis added).[27] In fact, the overall tenability actually improved along the intake pathway, increasing with proximity to the floor and the inlet.[28]

Past research and study have only focused on the effects of ventilation on the fire and the immediate area of involvement. The results from these experiments cast light on what was occurring in the rest of the structure, where firefighters are operating and approaching from, and more importantly, where potential victims will be located. Because the threshold of this positive impact varies substantially relative to the environmental factors and the fire conditions, it is of the utmost importance that ventilation be closely aligned with the progress of interior operations. Ventilation must occur in conjunction with suppression and isolation measures.

Tenability

The conditions within the fire building that determine the extent to which operations can be carried out, as well as the survivability of potential victims within a specific area, are known as *tenability*. The level of exposure and its subsequent impact are essentially based on the same three principal factors used for radiation incidents in hazardous materials operations: time, distance, and shielding. As with all things on the fireground, time is of the absolute essence. The longer the duration of the exposure to the fire and its by-products, the more detrimental the effects become. The distance aspect not only refers to the fire, but also to the floor. Because heat and smoke naturally rise, the higher up the position, the hotter and more toxic it becomes. With the gas concentrations typically posing a more imminent threat than the thermal exposure, victim elevation can be of greater significance to tenability than victim proximity to the fire.[29] The shielding component addresses the degree of physical separation from the fire and its by-products. Within the fire building, the most advantageous position is one that is fully isolated. The more pronounced the barrier is, the more it will resist fire spread, as well as limit heat saturation and the by-products from contaminating that enclosed area.

The most prominent misconception regarding ventilation is its impact on the conditions. A common belief within the fire service has been that even on its own, ventilation (especially vertical ventilation), can *globally* improve tenability. That assumption has been derived from the distinctive relief that can be experienced

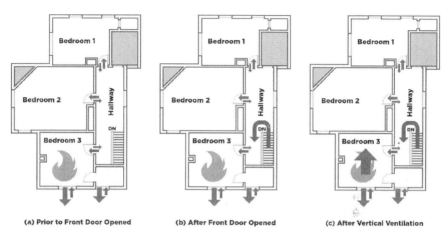

Figure 3–5. The changes in air flow following entry and vertical ventilation, illustrating the limited air exchange occurring within the rooms distal of the ventilation (intake) pathway.
Courtesy: UL-FSRI

at the floor level when ventilation first takes place. By creating an opening, especially one that is opposite/above the advancing crews, the fire and its by-products are allowed to exhaust from the building, away from their position. This can alter the thermal balance and enhance the air exchange between the inlet (typically the entry door) and the newly formed outlet. The sudden lift of the thermal balance within the ventilation pathway above the firefighters' level, allowing fresh air to intake below, is responsible for that reprieve from the heat and smoke.

In this case, an increase in visibility and the oxygen concentrations, as well as a decrease in the toxic gas levels and temperatures, can be initially experienced, cumulatively improving victim survivability. The experiments for the UL-FSRI coordinated fire attack study found that "ceiling temperatures in the areas located between the fire and the low-pressure vent began to increase, while temperatures at lower elevations in the same areas decreased as fresh air was entrained."[30] The temporary improvements in tenability as a result of the lifting effect, however, *only occur where the fresh air flows in toward the fire* (fig. 3–5).

The spaces distal of the intake pathway, even if directly connected by an entryway, will not be substantially affected. Unless those adjacent areas have their own inlet openings present to draw in fresh air, there will be no considerable air exchange or lift of the thermal balance within those spaces, causing the conditions and thus the tenability to remain largely stagnant.[31] While ventilation can provide a momentary benefit along the intake pathway, the additional supply of oxygen will accelerate the combustion process once it reaches the fire, especially given the underventilated conditions being experienced at the vast majority of today's fires. The increased output of the fire can quickly exceed the exhaust capacity of most openings created if they are not closely coordinated with effective control measures. The conditions and tenability outside the fire room can start deteriorating rapidly.

Once the opening created has been consumed, the fire will begin to spread laterally. This fact exposes another misconception regarding ventilation's ability to confine the fire and limit its horizontal spread. Because most fires have become so fuel rich, ventilation *alone* cannot practically exhaust their potential output. Any additional oxygen provided leaning out the environment will cause the fire to further develop. Venting the fire until it achieves peak combustion and becomes regulated by the fuel load, thus transitioning it to a fuel-limited state, would require an opening that is likely not even possible in the appropriate time frame.[32] The typical ventilation openings that are made, both horizontal and vertical alike, are not sustainable when executed preemptively. If water is not swiftly applied to the seat of the fire or the affected area is not isolated, any positive effects of ventilation on the conditions and tenability will be fleeting.

On the flip side, considerable detriments to tenability, specifically victim survivability, can be experienced if ventilation is unnecessarily withheld. While restricting ventilation will certainly prevent any additional fire growth, if the by-products of combustion are not promptly released, the exposure to any victims will be proportionately increased as well. The study also found that "the experiments in which toxic gas concentrations remained highest for the longest, were those in which no timely ventilation actions were performed close to the occupant location."[33] In turn, ventilation must be well coordinated with interior operations to limit this time frame.

Our interventions must be focused on rapidly improving the conditions *throughout the structure*, including any remote areas that do not receive immediate relief from the openings created to gain access and support extinguishment (fig. 3–6). Opening the entry door and venting the fire room, even from above, will initially provide little benefit for any victims located outside of the intake pathway between the entry door and the fire room.

This concern was supported by the coordinated fire attack study, where in some experiments it took *more than two minutes* to show improvement in remote occupant packages. Even when the fire room was vertically vented after interior suppression, "although conditions in the hallway and fire rooms quickly improved, the gas concentrations and resultant toxic dose in the remote bedroom did not improve as quickly."[34] Rapid occupation of the interior must be a top priority to access, search, and ventilate these remote spaces, ensuring they are isolated first if ahead of extinguishment. This local, horizontal ventilation can improve gas concentrations and visibility and reduce thermal insult at the floor level within the first minute.[35]

3 • Ventilation Defined   33

Figure 3–6. A large volume of dense smoke pours out of the room after the window is vented, relieving conditions inside.
Courtesy: Matt Daly

OBJECTIVES AND METHODS

To ventilate efficiently, a firefighter should know why it is done, as well as how, when, and where....The selection of means, or how to ventilate, is determined by the goal or minor objective to be attained in view of the major objective.

—Deputy Chief Charles Walsh,
FDNY

The commonly utilized phrases, "vent for fire" and "vent for life," intended to describe the two primary ventilation objectives, have been more aptly rephrased as "vent for extinguishment" and "vent for search," respectively. This alternative not only emphasizes the supportive and task-oriented nature of ventilation, but hopefully also dispels any associated misconceptions regarding its capabilities and the resultant effects, reinforcing the need for coordination. For that very reason, these semantical revisions were adopted by the FDNY within their operational doctrine in 2013, following the release of the UL-FSRI ventilation studies.[1] A third objective, "vent for access," has been added to this book, emphasizing the association between entry and ventilation. Any opening made, even if done passively when accessing the building, will provide ventilation, creating new ventilation pathways and prompting a change in the conditions.

Venting for Extinguishment

Widely known as "venting for fire," the intended objective of venting for extinguishment is to facilitate the fire attack efforts by providing an outlet for the fire and its by-products opposite and/or above the advancing engine company as they move in for suppression. This phrase can be slightly misleading, however, and may perpetuate the belief that preemptive ventilation of the involved area can *sustainably* confine the fire and *globally* improve conditions. The success of ventilation hinges on its close coordination with the onset of extinguishment. Properly timing the creation of an appropriately sized and placed opening will enhance the air exchange. The resulting ventilation will provide physical relief and will improve visibility along the intake pathway, in addition to aiding in the confinement of the fire, as the nozzle team makes their push and knocks down the fire (fig. 4–1).

Figure 4–1. Firefighters louver the cut section of the roof and punch down the ceiling to vent the spaces below.
Courtesy: Alan Chaniewski

Venting for Search

Also known as "venting for life," the purpose of venting for search is to support search and rescue operations, as well as improve tenability. The phrase "venting for life," however, is fairly ambiguous and provides no specific tactical reference since all forms of ventilation should be centered on life safety and enhancing victim survivability. Once a space (such as a stairwell or a remote bedroom) has been isolated and/or the fire is in check by the attacking handline, taking and opening windows, skylights, or bulkheads will exhaust the pent-up heat and smoke and provide relief for both the victims and the firefighters searching for them. This will increase visibility and more importantly will improve the tenability of the affected area. When there is *confirmed entrapment*, however, this form of ventilation can be executed to initiate a targeted search and rescue operation (*vent-enter-search [VES]*), despite the progress of the fire attack or the status/ability of door control (isolating the space). Although there is the inherent risk of accelerating fire growth and spread directly toward the newly created point of low pressure, it is far outweighed by the potential of saving a life (fig. 4–2).

Venting for Access

Although often overlooked, venting for access is the most basic objective of gaining entry to the building or to an enclosed area within it. Whenever any opening is made, it will serve as a point of ventilation. Just like breaking a window or cutting a hole in the roof, opening a door will impact the fire and the conditions whenever there is a direct pathway (fig. 4–3). This also applies to breaching ceilings and walls to access void spaces or the attic/cockloft area

Figure 4–2. A firefighter vents the window and prepares to make entry.
Courtesy: Tim Olk

Figure 4–3. Smoke pushes from around the seams of the doorframe as firefighters work to force entry.
Courtesy: Bob Pressler

to expose hidden fire. Whether one is passing through a door to make entry or opening up to overhaul the concealed spaces, such actions will result in a form of ventilation and must be treated as such. The fire cannot differentiate between the intended functions of the openings; it just reacts. An opening is an opening.

Vertical Ventilation

After evaluating the nature of fire behavior and air movement, ventilation is clearly most effective when conducted via vertical channels. Because gases expand and rise as they are heated, moving from areas of high pressure to low pressure, the by-products of combustion are most rapidly exhausted via an opening above the area of involvement. Venting above the fire (the area of the biggest pressure differential) will leverage the buoyancy of the gases to redirect the heat and smoke from below.[2] Serving as a dedicated outlet, vertical ventilation maximizes the output and lift of the thermal balance, producing a unidirectional exhaust that improves visibility and provides relief along the intake pathway. When properly timed with fire attack, the impact of topside ventilation can make it the preferred exhaust opening for the crew making the push, but more importantly, for any victims who are trapped wherever the fresh air is being entrained.[3]

> The fire cannot differentiate between the intended functions of the openings; it just reacts. An opening is an opening.

While the opening is traditionally placed directly over the seat of the fire, strong consideration should be given to cutting the vent hole just outside the threshold of the fire room when the fire is on the floor directly below the attic/cockloft. Because of its efficient air exchange, vertically venting directly over the fire will cause it to grow at a greater rate than if the hole is created slightly remote from the fire. Furthermore, by capturing the adjoining hallway/living space, that area will be vented as well. Although the fire will be drawn out of the room through the doorway, only the temperatures in the top few feet (known as the *upper register*) will be increased when coordinated with effective suppression, as the fire promptly exhausts out of the vent hole immediately above. The experiments for the UL-FSRI coordinated fire attack study found that cutting a vertical ventilation hole in this manner caused gas concentrations and temperatures along the intake pathway to decrease close to the floor. "Locations closest to the inlet portion of the flow path [received] the most immediate positive benefits, decreasing the CO and CO_2 and increasing the O_2 at that location. Additionally, visibility improved in the flow path between the front door and the fire room," cumulatively improving tenability within that space.[4]

That being said, however, things are not always so cut and dried. There are many variables and complications that must be considered before committing to vertical ventilation. First and foremost is the location of the fire, which is the most important deciding factor when venting for extinguishment. Vertical ventilation in this case is most successful when the fire is in the attic and/or knee-wall space, but it can also be applicable for fires on the floor directly below (fig. 4–4). Creating a vertical ventilation opening that has a direct pathway to the area above the fire will ensure the heat and smoke are exhausted in the quickest manner, making it "an ideal opening."[5] Doing so may be as simple as clearing out a skylight or as involved as cutting a hole in the roof and punching through the ceiling of the floor below.

Vertically venting the floor directly beneath the attic/cockloft area poses some inherent challenges and drawbacks. If the attic/cockloft area is not involved, cutting a hole in the roof and penetrating the ceiling below will bring the smoke and fire through that previously unaffected space, further emphasizing the

Figure 4–4. Ladder 1 Firefighter Pete Pyzik opens the roof to vent the attic and knee-wall void spaces.
Courtesy: NBFD

importance of timing. If the fire has not communicated to the attic/cockloft area, vertical ventilation will bring it there, *escalating into a structure fire if it is not promptly followed by extinguishment.*[6] Furthermore, the opening made in the ceiling may not match the size of the hole cut into the roof, reducing its exhaust capacity. As soon as the connection is made between the two spaces, the fire will want to begin venting out. Be sure to exercise caution when operating over the hole, as the fire can suddenly and forcefully vent out.[7] The rate at which it does so is conditional and will dictate the degree to which you are able to clear the ceiling below (fig. 4–5).

Creating an opening of comparable size within that time can be problematic when dealing with plaster and lath, especially when metal/mesh sheets are used in place of or in addition to traditional wood furring strips. In the worst-case scenario, the ceiling may not be breached at all due to the presence of decking (typically plywood) placed on top of the joists to create a walking or storage surface in the attic. Cocklofts/attics could potentially house a large quantity of various combustibles.

Vertically venting for search, however, is typically far less complicated and manpower/labor intensive, specifically in regard to flat-roof multiple dwellings. This form of ventilation is predominately intended to clear the enclosed stairwells

Figure 4–5. Firefighters contend with flames as they cut the roof to vent the attic and living space below.
Courtesy: Michael Carenza Jr.

of smoke to increase visibility and release toxic gases. This type of vertical ventilation facilitates interior operations and enhances the tenability for any victims trapped on upper floors or attempting to evacuate from them. The necessary ventilation can be accomplished by simply opening up a bulkhead door and/or the skylight over the stairwell. While these actions can be potentially lifesaving and should be initiated as soon as possible, *they must be coordinated with interior operations.* As long as the door to the fire floor and/or the fire apartment is controlled, the openings can be made prior to the onset of the fire attack.[8] With the stairwell effectively isolated, the resultant ventilation will not have an open pathway to the fire.

Preemptively venting the stairwell, especially if it is the stairwell from which the nozzle team is initiating their advance (i.e., the *fire attack stairwell*), may have to be temporarily withheld at fires in high-rise buildings or when a wind-impacted condition otherwise exists. Unless the access to the targeted stairwell can remain controlled throughout the nozzle team's approach and attack, creating these downwind vertical openings should be delayed until the team has the fire in check. Otherwise, this point of low pressure can rapidly intensify the growth and spread of the wind-driven fire.

A major factor that affects vertical ventilation operations is the building construction and its features. The chief concerns include the roof (pitch, sheathing, and structural members); height (dictating the required ladder size/type); access (setback and overhead wires); obstructions (solar and HVAC units); and security (fencing and steel plating). Another variable that must be accounted for is the weather. For areas that drop below freezing, snow and ice can be considerable hindrances during the winter months. Significant snow accumulation can create large drifts that may prevent aerial apparatus from getting into functional or optimal positions, as well as making it difficult for firefighters to maneuver both on the ground and on the roof. Moreover, the weight of the snow places an added live load on the roof, further stressing the structural members. These conditions also make traction and footing problematic, increasing the likelihood of a slip-and-fall event. The potential for such an event is particularly a concern with peaked-roof dwellings, which necessitate aerial device operations or the use of roof ladders to mitigate the hazard. These components play into the resource requirements, the duration of the operational period, and the ultimate effectiveness of the operation.

The more complicated or challenging the environment, the greater the demands placed on the personnel and the more lead time will be required to complete the task. This can increase the cumulative risk and reduce the probability of successfully coordinating the operation. The background and capabilities of the crews are of particular importance, as their level of experience and competency will determine how well they are able to overcome any obstacles. The following questions must be posed before selecting your tactical approach:

- Does your department have the resources to appropriately contend with the conditions present?
- Does your department frequently encounter these obstacles?
- Does your training incorporate these variables?
- How proficient are the members performing the operation?

When conditions are favorable for venting vertically, the ultimate deciding factor becomes the resources available. The vast majority of fire departments across the country are forced to operate with substandard staffing. There are many agencies that struggle to simultaneously complete the two primary functions of fire attack and search as independent operations. Venting the roof, therefore, can be detrimental to those priority operations *when it is not essential to achieving the intended objectives.* By misappropriating resources, the necessary personnel may not be available to properly get handlines into position and promptly complete the search, not to mention causing undue property damage.[9] If sufficient horizontal openings are present, a single firefighter with a hand tool (and a ground ladder) can adequately support these efforts.

It is important to note that while vertical ventilation may be a superior tactic in terms of yielding a more efficient air exchange, venting horizontally is often capable of producing the essential relief with less time, effort, and logistics, which can give it strategic superiority.[10] With that in mind, fireground commanders must diligently prioritize the tactical objectives. They must execute the interventions that will have the greatest impact on the strategic priorities, based on the conditions and the resources at their disposal, to maximize the return on investment[11] (fig. 4–6).

Figure 4–6. Firefighters strip off the terra cotta tiles to expose and cut the roof decking.
Courtesy: Erik Haak

Horizontal Ventilation

While venting horizontally may not have the exhaust efficiency of its vertical counterpart, its ability to facilitate extinguishment and search can be at least comparable. As long as the attic/cockloft is not significantly involved, horizontal ventilation will typically yield the desired effect.[12] For most contents/compartment fires, simply taking the windows can provide the necessary relief for the engine company making the attack. When comparing the area of the openings created by both tactics, specifically on peaked-roof dwellings, window venting may meet or even exceed the exhaust capacity of a hole cut into the roof, especially on steeper pitches.

The reason for this anomaly is that only a single rafter is being rolled to create the roof opening. Often done unintentionally, this subconscious behavior or "training scar" may result from the use of roof props to train on this skill set. Due to the price of plywood and oriented strand board (OSB), as well as the logistics involved, holes of smaller dimensions are cut to maximize their use. Even when done out of necessity due to a steep roof pitch, if the single-rafter cut is not extended down at least an additional 2'—accounting for the decreased width (only 28")—the standard 4' × 4' (16 ft^2) hole would be reduced to just under 10 ft^2. The loss of more than one-third of the opening's area would drop the heat release rate capacity from approximately 3 MW (2,850 Btu/sec) to 1.9 MW (1,800 Btu/sec). The average residential space, however, typically has at least two windows (roughly 10 ft^2 each). With a total area of 20 ft^2, the horizontal openings have a capacity of roughly 2 MW (1,900 Btu/sec), yielding a slightly greater output.[13]

Figure 4–7. The gable louver is not enough to vent the fire, requiring that a hole be cut.
Courtesy: Dave McCabe

While windows largely function as both inlets and outlets, producing a less efficient bidirectional flow, their exhaust output will be enhanced when the attacking handline begins flowing, which will increase air entrainment and create its own pressure front. Additionally, the exhaust efficiency of the windows will increase as soon as the front door is opened, prompted by the additional intake of fresh air.

When the fire is on an upper floor and an open stairway is present (as in a private dwelling), the front door will be situated beneath the fire floor (also known as "*below level*"), serving as a pure inlet and further increasing the rate of air exchange.[14] In this instance, window venting may be a more prudent tactic than cutting the roof and pushing down the ceiling below. In these circumstances, window ventilation can effectively support the intended objectives and maximize the overall efficiency of the operation.[15]

Even when the fire is within the attic space at a peaked-roof dwelling, the natural openings present *may* be enough to do the job, depending on the conditions encountered. FDNY's manual on private-dwelling operations, *Ladders 4: Operations at Private Dwellings*, states, "Physically cutting and opening the roof is usually not considered an initial operation at peaked-roof dwelling fires. The venting of attic windows or louvers is frequently sufficient for ventilation purposes" (fig. 4–8).[16] This is not always the case, however, as the availability/capacity of those openings may not be conducive to such an approach, based on the building construction, the fuel load, and the level of involvement (fig. 4–7).

This fact is especially true in cases of large, walkable attics, which are commonly used for storage or as additional living spaces, and can drastically increase

Figure 4–8. Fire exhausts from the vented louver.
Courtesy: Patrick Dooley

the fuel load and potentially altering the layout (compartmentalizing the space). If knee walls are present and the fire has entered into the voids, *horizontal openings will not ventilate those spaces*, allowing fire to spread and superheated gases to dangerously accumulate. Opening up the roof, especially in those cases, may be the deciding factor in successfully attacking an advanced fire in the attic/cockloft.[17] Additionally, if signs of a potential backdraft/cockloft explosion are present, vertical ventilation can also minimize the effects of an impending explosive ignition (known as a *deflagration*) or even prevent it from occurring on accessing the involved area.[18] Selecting the appropriate avenues to ventilate should be made on a case-by-case basis, with the conditions and resources dictating actions.

Horizontally venting for search, on the other hand, is greatly effective and should be given high tactical priority. For rooms remote of the intake pathway, performing local ventilation by taking the windows—when coordinated with extinguishment or isolation of the affected area—greatly improves the conditions within that space. The conditions may otherwise remain stagnant. The experiments for the UL-FSRI coordinated fire attack study found that even vertical ventilation over a hallway or a living space will not have a significant effect on the air exchange in the areas outside of the intake pathway. The results did, however, show that "when horizontal ventilation was performed at any available opening, gas concentrations, visibility, and skin burn assessment package temperatures dropped at a more rapid rate than in experiments where no local ventilation was performed," improving with proximity to the floor and to the inlets.[19] With uninvolved bedrooms most commonly falling into this remote area category, committing to aggressive searches is paramount. We must occupy and vent these spaces as soon as possible to more effectively and efficiently enhance their tenability and remove any victims. Doing so could provide them with the greatest chance of survival, especially when manpower is limited.

As with any form of ventilation, taking windows (especially those remote from the fire, must be coordinated) as creating a remote outlet will draw the fire to that new point of low pressure. Ventilation tactics should be aimed at establishing intake pathways through any area where victims may be trapped to enhance tenability of that space. As the study noted, "In many cases, the most effective way may be the *coordinated* horizontal ventilation of as many windows as necessary."[20] Doing so must be in cadence with extinguishment unless the space being ventilated can be isolated from the fire first, essentially the same principle of the vent/enter/isolate/search (VEIS) tactic but initiated from the interior. Simply closing the door behind you restricts the airflow pathway to the fire, allowing the windows to be vented on reaching the outside wall during the search ("venting as you go"). Venting horizontally, however, is not immune to complications or challenges. In vacant buildings, windows are predominately sealed up for added security. Traditionally done with plywood or OSB, this practice has evolved, especially in urban areas, to thwart potential intruders/vagrants. Recently, elaborate metal coverings called *vacant property security (VPS) systems* have been used to increase protection. Homemade versions, using plywood or OSB and 2' × 4' lumber or acrylic/plexiglass panels and angle-iron, fastened with carriage bolts, are more commonplace, often referred to as *HUD windows* because their use is approved by the US Department of Housing and Urban Development. Also present in high-crime areas are window bars, grates/mesh, and brick/block (masonry or glass) coverings. While these added security measures can be defeated with proper forcible entry techniques and equipment, it takes additional time and potentially more resources (fig. 4–9).

A similar condition exists in parts of the country that are subject to extreme weather, especially in the southeastern United States, where hurricane-rated windows are being installed along the coastal areas. This type of impact-resistant

Figure 4–9. Cutting window grates
Courtesy: Matt Daly

glass can also be found in other areas prone to storm activity, as well as high-rise, governmental, correctional, and high-value occupancies. Because of this robust construction, conventional ventilation methods may be rendered ineffective or at the very least inefficient, requiring more rigorous efforts and additional equipment and time. Another factor that greatly limits the use of horizontal ventilation is wind. If a notable wind condition is present, the openings being impacted (the upwind side of the building) must be controlled, potentially limiting the ventilation options.[21]

Hydraulic Ventilation

This method of supplemental ventilation is highly effective and can be implemented immediately following the fire attack, as it requires no additional equipment/personnel. By simply directing the stream from the handline's nozzle out an opening in the fire room (typically a window), the air entrained by the stream will create a negative pressure, drawing the smoke out of the building in a unidirectional flow.[22] While a fog nozzle, set to a narrow-fog stream, is the most efficient nozzle for this application, generating more than 8,000 cubic feet per minute (cfm) (at 150 gallons per minute [gpm] flow), aggressively operating the solid stream of a smoothbore nozzle in an O-pattern within the window opening "entrains a similar magnitude of air;" roughly 5,000 cfm (at 150 gpm flow)[23] (fig. 4–10). The air entrainment of a narrow fog stream can also be enhanced by

Figure 4–10. A smoothbore nozzle being used to hydraulically ventilate

manipulating it in that same pattern, producing upwards of 12,000 cfm (at 150 gpm flow), rivaling the output of some mechanical positive-pressure ventilation fans.

It may be possible to further improve the effectiveness of the smoothbore nozzle by removing the tip and dialing the bale back enough to disperse the stream and create a wider pattern, with the bale positioned at about halfway open. With the ball valve occluding a large portion of the nozzle body's wider orifice, the water velocity is increased in accordance with Bernoulli's principle, much like placing your thumb over the end of a garden hose. The slug of water traveling through the waterway is then broken up into smaller droplets, likely a result of the position of the ball valve restricting the waterway and its close proximity to (and the characteristics of) the terminal orifice, which is the 1½", male-threaded end of the nozzle body. While this technique has not been studied and analyzed scientifically, it has been anecdotally tested in the field with positive results (fig. 4–11).

In any case, the nozzle must be positioned back as far from the window as possible so that the stream envelops the greatest amount of the opening. Increasing the length of the stream yields greater air entrainment as well. Be sure to monitor conditions throughout the operation and shut down the nozzle as soon as the desired effect has been achieved. If the tip of the smoothbore has been removed (and hopefully safely stowed in one of the nozzle firefighter's turnout gear pockets), immediately replace it to address any remaining hot spots and complete overhaul, in addition to preventing the nozzle tip from being forgotten or misplaced.

Hydraulic ventilation is often a superior choice for postextinguishment ventilation for several reasons. The application of this method is executed by the nozzle team as soon as knockdown has been achieved. This makes the operation inherently coordinated and does not necessitate any other resources, a critical factor for agencies with limited staffing. Furthermore, hydraulic ventilation is achieved through negative pressure, resulting in less large-scale mixing of the

Figure 4–11. A stationary, narrow fog stream hydraulically venting
Courtesy: UL-FSRI

environment when compared to positive pressure ventilation and maximizing the flow/exchange of air. The air that is entrained behind the stream, largely through the entry door, is able to remove smoke from adjacent spaces within the intake pathway that have been contaminated. This includes common egress paths outside of the fire compartment in multiple dwellings. To maximize ventilation of common hallways and stairways, ensure that the entry door is fully open and control all other intake openings within the fire apartment, creating a greater draw at the door. Conversely, if the focus is on ventilating the fire apartment, creating local horizontal openings (i.e., windows) will increase the draw within it.[24]

Mechanical Ventilation

Mechanical ventilation employs gasoline-, electric-, or battery-powered fans to create positive or negative pressure within the structure. Electric box fans, commonly known as *smoke ejectors*, are typically placed within an open window with the fan directed outward. This creates a negative pressure within the space, drawing out the smoke in a manner similar to hydraulic ventilation. As with any other form of forced-air ventilation when supported by appropriately proportioned inlets and outlets, the greater the air entrainment, the more efficient the air flow will be.[25] These fans can also be stacked and used in tandem for doorways or large windows. An adjustable bar can be placed within the frame of the opening to hang the fan. It is important to ensure the fan is fitted to the opening to maximize the stream of air being produced. Additionally, it will be necessary to remove any window treatments that may block the intake side or get sucked into the fan blades, potentially causing damage.

When utilizing fans for postsuppression, positive-pressure ventilation (PPV) within the area of involvement, it is imperative that the fire be fully extinguished. This is especially true in buildings with combustible void spaces, most notably balloon-frame and ordinary construction, where fire may be present. These powerful fans, particularly the gasoline-powered units, will drastically accelerate the air flow, potentially in excess of 20,000 cfm.[26] The use of PPV will, as its name suggests, pressurize the space, forcing air throughout the structure wherever an open pathway is present. This influx of fresh air can enter into the void spaces and promote the development and spread of concealed fire.

The other issue posed by pressurizing the space is the potential degree of turbulence that can be created within it. When a high volume of air is forced into a structure in which the distal outlet is not sufficient, large-scale mixing of the environment can result from the overpressurization, disrupting the flow and limiting the exhaust potential. To maximize the effectiveness of PPV, the outlets should be greater than the intake (up to five times the size, depending on the output of the fan), increasing the air flow/exchange within the space and improving conditions more rapidly.[27] The fire's by-products can also be unknowingly driven into and collect within adjoining compartments if uncontrolled openings are present. The buildup of smoke and toxic gases (namely CO) can be problematic for large buildings and even deadly in multiple dwellings if occupants are being sheltered in place. CO can also be an issue when using gasoline-powered PPV fans if the engine's exhaust is not directed away from the building to a safe location.

Another use for PPV fans is to pressurize enclosed stairwells to prevent smoke contamination and preserve tenability to facilitate occupant egress, as well as enhance firefighting operations. This is typically employed at fires in high-rise buildings. Whichever method of mechanical ventilation is being selected, the opening being used by the fan should capture the full stream of air to maximize its output and the efficiency of the air exchange. These devices are capable of moving tremendous amounts of air and can be highly beneficial when used prudently. As with any other form of ventilation, however, mechanical ventilation must be conducted in coordination with the interior operations.

Positive Reinforcement

While on the apparatus floor, drilling on handline stretches, we received an alarm for a reported structure fire at 38 Broad Street, only a few blocks from the fire house. As we pulled off the ramp and rounded the corner onto Washington Street, we could see a large column of thick, dark smoke on the horizon. The large, four-story, ordinary-constructed, mixed occupancy (residential over commercial) building was located on the corner of Washington Street. As we crossed over Broad Street and pulled past, leaving room for the ladder company, we were granted a view of the rear. Fire was venting out the back bedroom window of the delta-side apartment and autoexposing to the rear porch, prompting my immediate transmission of a second alarm.

As we dismounted the apparatus, several bystanders had gathered and alerted us to a victim who was trapped on the top-floor landing, the occupant of the involved apartment. The victim was hanging his head over the railing on the bravo side, attempting to get relief from the intense heat and smoke. The fire had enveloped the delta side of the porch where the stairs were located, cutting off his means of escape. With the fire rapidly spreading, and the conditions deteriorating, the victim was in imminent danger (fig. 4–12).

As the first-arriving engine company, our priority was to protect the life hazard. We deployed a 1¾" handline up the rear porch to cut off the fire extension

Figure 4–12. Arrival conditions in the rear
Courtesy: NBFD

and restore the egress path. Traversing three flights of long, switchback stairs would prove to be a rather challenging (350') stretch. Ladder 1, running out of the same quarters, arrived on scene just moments later, setting up on the bravo/charlie corner. The Ladder 1 inside crew went up the front stairs to perform the primary search of the fire apartment. Recognizing the urgency of the situation, the Ladder 1 outside crew deviated from their standard roof assignment of performing recon and ventilation and went into rescue mode (fig. 4–13).

Figure 4–13. Smoke and vent-point ignition indicate a well-advanced fire.
Courtesy: NBFD

As we made the push and knocked down the porch, flowing and moving up from the stairs below, the Ladder 1 outside crew arrived in the bucket. While we were making the turn into the rear hallway, the victim was picked up by the outside vent firefighter and brought down in the bucket of the ladder tower. Because of the quick thinking and proficiency of the Ladder 1 outside crew, the victim only suffered minor smoke inhalation.

While the victim was being turned over to New Britain EMS, we entered the apartment. We were greeted by almost zero visibility and fire rolling out of the back bedroom into the kitchen, despite smoke being vented out of two windows. As the Ladder 1 inside crew was searching toward the front of the unit, we made the final turn. With exhaust openings opposite our advance, we easily pushed the fire back and quickly achieved knockdown. While we were extinguishing the fire, the second-due ladder company, Ladder 2, cut the roof and punched down the ceiling, enhancing visibility and providing additional relief. To expedite the process, my nozzle firefighter removed the tip from the smoothbore nozzle, cracked the bale, and began hydraulically venting out one of the windows. After the smoke lifted, it was confirmed that the fire had not extended into the cockloft (verified by the thermal imaging camera), and the incident was placed under control. Due to proper assessment of the situation and correct prioritization of the tactical objectives, the overall strategic mission was accomplished.

5

SIZE-UP AND DECISION-MAKING

No environment is ever static. As the environment around us changes, developing the situation allows us to maintain our most prized freedom of choice—to adapt our thinking and decision-making accordingly.

—Colonel Pete Blaber,
Delta Force

The RADE Loop

The *RADE loop*, a fusion of US Air Force Colonel John Boyd's OODA loop and London Fire Brigade Crew Commander Paul Grimwood's tactical ventilation strategy, is a model that outlines the decision-making process and the contributing factors. Its core elements are as follows: 1) Read the conditions; 2) Anticipate the progression; 3) Determine the needs; and 4) Execute the tactics.[1] The RADE loop is *not* a mnemonic to be recalled out in the field and used as a checklist or a sequential action plan. Rather, it was constructed to demonstrate how fireground decisions are made, identifying the influential factors and the information that is processed (largely subconsciously), as well as the relationship between these different facets. When we understand what is taking place and how all of the pieces are connected, we can more accurately assess a situation, identify the available options, and consistently and rapidly make better decisions (fig. 5–1).

Read the conditions

While it technically begins on the receipt of the alarm, the physical process is not initiated until arrival on scene. When pulling up to the fire building, the first step

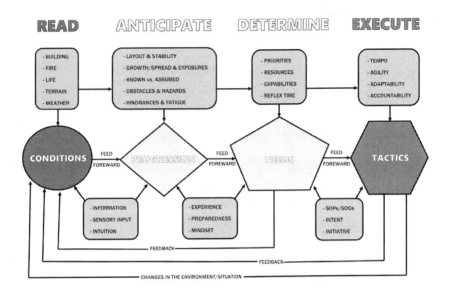

Figure 5–1. The RADE loop is a decision-making model that identifies the contributing factors and variables, outlining the continuous and dynamic nature of the process.
Courtesy: Nicholas Papa

is to immediately read the conditions. Start with the building, identifying the construction, height, roof style/pitch, openings, and security measures. Evaluate the presentation of the smoke (volume, velocity, density, and color) and fire to determine its location and extent. Doing so will also aid in recognizing the inlets and outlets, and identifying the existing points of ventilation as well as their current and potential pathways (known as the *ventilation profile*) (fig. 5–2).[2] Gauge the life hazard based on the dispatch information, the time of day, the occupancy type, and any reports from occupants or bystanders. Assess the terrain to identify any conditions that could hinder operations: long setbacks, graded lots (i.e., walk-out basements or cellars), overhead wires, and fences, gates, and walls. The weather (most notably the wind velocity) can significantly impact the fireground.

Anticipate the progression

The information gleaned will enhance your ability to predict how the fire will react so you can anticipate the progression of the conditions. Forecasting the fire's rate of growth, spread, and direction of travel, as well as the layout and structural integrity of the building, is imperative to calculating the operational window and formulating an action plan. Often called "BAG-ing" the fire, locate where it has *been*, where it *is*, and where it is *going* (fig. 5–3).[3] Differentiate from

Figure 5–2. The smoke exhausting from the dormer indicates a growing attic fire.
Courtesy: Nicholas Papa

the standard assumed life hazard versus one that is known, in addition to the last known location of any trapped victims, as this will likely alter your tactical approach and the subsequent level of risk you are willing to take. This is crucial in locating the most advantageous area of operation to aid in the confinement of the fire and protection of the exposures, to maximize the survivable space, and to locate and remove any trapped victims in the most effective and timely manner. When observing the terrain and weather conditions, attempt to predict how the obstacles, hazards, and hindrances will impact the fire behavior and the firefighting operations, including apparatus positioning, accessibility, maneuvering, and traction, as well as the level and speed at which crews become fatigued and need relief/rehabilitation.

Determine the needs

Fireground commanders must account for and prioritize the operational demands of the incident and weigh the tactical options against the available resources. Once they determine the needs, they must identify how they can meet those objectives based on the present capabilities and the subsequent reflex time required to do so.

Figure 5–3. The smoke condition from the front dormer window is rapidly worsening.
Courtesy: Nicholas Papa

Execute the tactics

Once the decision has been made, the plan must be communicated properly. Providing the *commander's intent*—informing subordinates what is to be accomplished and for what purpose—is the key to ensuring that the tactics are executed appropriately and the desired end-state is achieved.[4] The command structure, however, should be decentralized to provide the crews on the front line with the authority to make tactical-level decisions to best mitigate the situation that they are being confronted with, based on their firsthand knowledge of the conditions.[5] While operations should adhere to any established processes (standard operating procedures and guidelines) to ensure everyone is operating out of the same playbook, there must be latitude for discretion. It must be understood that when composing and integrating these systems, "operational doctrine should not be a formula, but a roadmap for thinking that allows for decision-making."[6] Subordinates should be empowered to take initiative if an advantageous opportunity arises, even if it means deviating from the script, as long it is communicated and justified. Once an operation has been executed, the impact must then be gauged to determine its effectiveness and if any further action must be taken, essentially restarting the RADE loop in a continuous cycle.

The importance of an evolving size-up cannot be overstated. Being able to evaluate the situation accurately and concisely predict how it will progress is the very essence of situational awareness and the foundation for sound decision-making. FDNY Deputy Chief Tom Neary once said, "Size-up is what separates the good ones from the great ones." Each incident we face will present its own unique set of variables that must be specifically accounted for and addressed accordingly, especially regarding ventilation. The ventilation needs of an incident are conditional, and as a result, they are highly dynamic. In turn, ventilation must be tailored to each incident, as there is no stock approach or universal tactic. No matter which form of ventilation is utilized, it must be communicated to all members operating, coordinated with the interior crews, and controlled to limit the growth and spread of fire, facilitating firefighting operations and enhancing victim survivability.[7] Successful ventilation operations thus hinge on the presence of competent commanders and skilled operators (fig. 5–4).

Figure 5–4. The firefighter makes his way down from the roof after fire blowtorches out of the vent hole that he cut.
Courtesy: Nicholas Papa

Fireground Tempo

Accurately assessing the fire's rate of change is imperative to establishing the correct *operational tempo*, which is defined as "speed relative to a problem set."[8] The purpose of our tactics is to positively influence the fire and the environment—to "make it behave"—so we can gain the upper hand. *Fireground tempo* is defined as "a series of disciplined, appropriate and timely tactical actions that promote mission accomplishment and enhance survivability...performed at a rate quicker than the environment's [fire's] rate of change."[9] Achieving an advantageous tempo requires the ability to rapidly extract and process information from all forms of sensory input, but predominately visual, auditory, and tactile input. The more we understand about the fireground and our tactics, the more efficiently and accurately we will be able to size up an incident, and the more effective our intuitions will be.[10]

The ability to establish tempo (i.e., set the pace) is often determined by one's level of situational awareness and intuitive decision-making ability, recognizing emerging cues and trends to address any threats and capitalize on opportunities. *Intuition* is the capacity to readily match these patterns to past events or previous knowledge stored in our memory.[11] While this is largely developed through firsthand experience, it can be honed through realistic training and real-world reinforcement, with quality repetition as the common denominator. *Performance*, therefore, is a matter of our individual preparedness and mindset.[12] Those operating at a high level can anticipate the progression and determine the needs subconsciously.[13] Because of this rapid cognition, they can seemingly bypass those intermediate steps and devise an appropriate "action script," immediately executing the tactics.[14] In turn, the four elements of the RADE loop are color-coded in the same fashion as a traffic light, representing the need for stopping to "read the conditions," taking caution to anticipate the progression and determine the needs, and executing the tactics (*going*).

Speed is one of the primary drivers of establishing tempo and is critical to gaining what US Navy Admiral William McRaven, Commander of the Special Operations Command, refers to as "relative superiority: decisive advantage over a larger or well-defended enemy."[15] Gauging how long it will take to get in position and complete an operation based on our resources and capabilities (called *reflex time*) is critical to determining how the incident priorities can most effectively and efficiently be addressed. This requires that the incident commanders and company officers be well aware of the strengths and weaknesses of both their crews and their equipment, understanding their limitations. Gaining advantage is more than just speed. It requires careful timing and executing at the pivotal moment. Gaining a sense for the rate at which the conditions are changing, along with awareness of the progress of the crews, will be essential to establishing the correct operational tempo and exploiting every potential opportunity that arises. "Timing means knowing when to act and, equally important, when not to act."[16] When it comes to the matter of ventilation, no truer statement could be made (fig. 5–5).

Establishing a superior pace goes beyond the initiation and completion of a single task. It is essential for the entire operation as a whole. Implementing the initial operations is often the easiest to achieve in this regard, particularly from a command and control aspect. The first arriving units, especially when strong operational systems are in place, go right to work addressing the primary objectives. Challenges may arise, however, when the incident is more complex or severe and these crews need immediate assistance or replacement. The speed with which they are able to make these transitions (known as *agility*) is imperative to seizing and maintaining operational momentum. Along those same lines is the capacity to adjust to changes in the environment or the situation at hand (known as

Figure 5–5. The smoke and vent-point ignition indicate a well-advanced fire.
Courtesy: NBFD

adaptability). The rate at which we do so is directly proportionate to our capacity for anticipating and improvising as the fireground evolves.[17] Even so, it is far more advantageous to "shape the environment, rather than adapting to it."[18] By staying alert and maintaining optimum readiness, we can better recognize emerging signs and have the flexibility to take proactive measures and combat the developing situation (fig. 5–6).

As we operate, notably in the performance of ventilation operations, we must ensure that we outmaneuver the fire, *not each other*. Doing so is achieved through disciplined and skillful operations and maintaining accountability at all levels. We must understand the impact and limitations of our ventilation tactics and acknowledge that we cannot practically "out-vent" today's fuel-rich fires. Ventilation must be accompanied by the timely application of water on the seat of the fire. Ventilation should be prompt but not premature, in cadence with the advancing engine company and the progress of the search crews. There is a narrow window of opportunity when the tactic is executed prior to extinguishment (unless the

Figure 5–6. The dormer window was taken after the hole cut was consumed.
Courtesy: Nicholas Papa

Figure 5–7. The IC monitors the radio traffic and observes the conditions.
Courtesy: Nicholas Papa

vented area is isolated from the fire), which is further complicated by its variability (fig. 5–7).

Venting as the engine company is moving in just before the onset of suppression carries the inherent risk of rapid fire growth and spread. Nevertheless, it must also be understood that "prompt ventilation will result in the best outcomes for potentially trapped occupants."[19] Determining the proper location, amount, and most importantly, the timing of ventilation is essential to accomplishing this task, ensuring it achieves its intended effect. We must also acknowledge and appreciate that "tempo is itself a weapon—often the most important. Superior speed allows us to dictate the terms and is necessary in order to concentrate superior strength at the decisive point."[20]

Incident commanders must ensure that they have ample resources to mitigate the incident effectively and efficiently while also accounting for any likely contingencies. They must plan not only for what they need now, but for what they may potentially need as the situation progresses. This is especially true during extreme weather situations, where crews will experience greater levels of fatigue and will need to be relieved at shorter intervals. Doing so requires a ready reserve to be established and maintained until the incident is placed under control. By "stacking the deck," there will be a surplus of personnel on standby to tap into at a moment's notice. These assets can be mobilized at any point to immediately support or replace the initial units operating, as well as to assist or backfill the rapid intervention team (RIT), should they be activated.[21] For many agencies, this will require mutual-aid or automatic-aid agreements. This is not a sign of weakness but of accepting reality, and more importantly, responsibility.

Unfortunately, the failure to implement these agreements and request additional resources is a fairly common occurrence for some organizations, and a failure to do so is becoming an even greater detriment in the wake of the continual trend toward declining staff levels. This issue is typically fueled by pride and ego, especially when it requires assistance from other agencies. Sometimes there may be apprehension about stripping the community of its remaining resources and not being adequately prepared in the event of a simultaneous incident. Commanders must remember that the incident that they are *presently* in charge of is their *sole* priority, and they are ultimately responsible to ensure that it is mitigated appropriately. The last thing you want is to look in the street and have no one left when you need them the most. Like the old saying goes, "A bird in the hand is worth two in the bush."

Learning the Hard Way

In the late evening hours of an early winter night, we received an alarm for a report of "black smoke coming from the roof" of the building at 42 Connerton Street. Given the time of year, this might have seemed like another good intent call, a result of a delayed ignition. Typically, however, those incidents are accompanied by a report of black smoke coming from the chimney, *not the roof*. On reading the "rip-and-run," we noted that the call originated from 24 Connerton Street, three buildings down and to the west of the incident building. When potential fires are reported as "next-door to" or "across the street from" an address, there is a higher probability of a legitimate incident, especially after sundown, as the conditions required to be detected by a neighbor are often significant. This also increases the likelihood of entrapment, as the fire is not being reported by the occupants themselves, who may be sleeping or otherwise unaware.

Engine 5 was the first to arrive and gave the initial on-scene report of "smoke coming from the roof." Several seconds later, a follow-up transmission was made, declaring, "Attention all companies, working fire," with no true size-up provided for the incoming units. As the second-due engine company, we arrived less than a minute later and were tasked with establishing a positive water supply for Engine 5. With a hydrant located directly across the street, we beached our apparatus on the curb, several houses back, and walked up to hand-jack the supply line off of Engine 5's apparatus. As we approached, a large volume of smoke was seeping from the entire roofline of a three-story, ordinary multiple dwelling (a six-family tenement) and hanging low in the street (fig. 5-8).

As we grabbed the hose and equipment from the apparatus, the Engine 5 nozzle firefighter was standing by on the tailboard, with a 1¾" handline loaded on his shoulder. The officer, on the other hand, was entering through the front door, not scouting ahead but rather attempting to alert the occupants, as he saw no indication that anyone had evacuated the building. What he was not aware of, however, was that the residents had been displaced the previous day, as the building had been condemned by the fire marshal's office and the building department after an incident involving the utilities. This action not only delayed the first handline from getting into operation, but also prevented the completion of a 360-degree size-up, which would have identified the location of the fire and the best point of access. This would be the first domino to fall.

Figure 5-8. Smoke condition shortly after arrival
Courtesy: Patrick Dooley

In roughly a minute's time, we made the hydrant and supplied Engine 5, completing our primary objective. As I was walking over to the incident commander to obtain an assignment, the Ladder 1 officer reported that the fire was located in the basement, as smoke and fire were visible through the delta-side windows. With no updates from Engine 5 on their location or progress, we were ordered to stretch a handline to the basement. We made our way back over to Engine 5's apparatus and stretched a 1¾" handline down the delta-side alley to the entrance of the enclosed rear porch. At the doorway, we flaked out and charged the line and began masking up, observing a large body of smoke exhausting out of the entire opening, which greatly obscured visibility. With the line charged, we moved in and began to make our way down the stairs. Only a few steps into our descent, we ran into Engine 5 *camped out on the stairs with an uncharged handline*. Not only were they in a dangerous position—stationary in the exhaust path of the fire and with no water—but because their status had been unknown, we were unnecessarily deployed to that location. This misallocation of resources and duplication of effort put us further behind the curve.

> As we operate, notably in the performance of ventilation operations, we must ensure that we outmaneuver the fire, *not each other.*

After trading places, we entered the basement, where we encountered zero visibility and fire in the overhead, running the joist bays and concentrated in the charlie/delta quadrant. As we were extinguishing the fire, we requested horizontal ventilation of the basement windows in that area. The fire in the basement was quickly knocked down, with the attack having been aided by a broken water pipe. I then made a transmission that the bulk of the fire was knocked down and that we had it contained to the basement, which could not have been further from reality. Being remotely located in the basement, my operational lens was extremely narrow. With my limited perspective further skewed by the lack of visibility, I did not have the intel to make such a broad-based statement. This misinterpretation initially caused some confusion and demonstrates why assumptions should not be made on the fireground, especially when providing radio reports.

Once we received ventilation, the smoke began to lift, and the conditions improved. After the majority of the smoke had been exhausted, we observed that the seat of the fire was in the same location as the building's utilities (the boilers and hot water heaters). With a significant level of char (also known as *alligatoring*) on the large floor joists, this raised a major red flag. With the pipe/utility chases running unobstructed through the center of ordinary-constructed buildings, vertical fire extension was a very real possibility. Around this time,

Figure 5–9. The smoke characteristics indicating the deterioration of conditions
Courtesy: Patrick Dooley

companies performing primary searches above began reporting fire in the walls, running the voids on floors one and two (fig. 5–9). The Ladder 1 outside crew had also observed embers spewing from the vent pipes on the roof, an indicator of vertical extension in the pipe/utility chase and most likely involving the cockloft. This critical piece of information, however, was not effectively communicated to the incident commander. As a result, the Ladder 1 outside crew was relocated to the interior to assist with opening up for the engine companies.

At this point, the fire began to break out *hard*. The Ladder 1 inside crew, operating on the first-floor, delta-side apartment, had heavy fire in the charlie/delta quadrant and also experienced a localized collapse of the kitchen floor due to the severely burnt floor joists from the basement fire below. The Engine 5 crew, located in the unit directly above, was combating similar conditions. The Engine 2 crew, which did not have a handline, was then ordered to the third floor to check for further extension. As they entered the delta-side apartment, they updated the incident commander that they had significant involvement in the cockloft above, along the delta-side. Given the concerning reports from the interior and the conditions from the street worsening, the incident commander began withdrawing crews from the building.

As the Engine 2 crew was making its way out of the apartment, large sections of plaster and lath from the ceiling began crashing down in the area. With no ventilation hole having been cut in the roof, these openings became the path of least resistance and allowed the heavy fire and pressure in the cockloft to move downward, bringing the heat and smoke to the floor. The sudden and complete loss of visibility and the spike in heat led to the members becoming disoriented and separated, prompting Engine 2's officer to call a mayday. At the time, we had just ascended the basement stairs in the rear porch. We immediately notified the incident commander of our position and that we were relocating to the third floor to assist. Our close proximity and knowledge of the location of the Engine 2 crew (based on the officer's previous update) would prove to be significant in the outcome. Other crews were either actively evacuating (and exchanging their SCBA air cylinders out in the street), while others were being reassigned as the operation was shifting to a defensive posture. This had momentarily left the position of rapid intervention team vacant and forced the first available crews to quickly fill this role and deploy with the necessary equipment.

As we made it to the porch landing, the smoke was boiling out of the third-floor rear entrance. We entered and moved down the hallway to the back door of the delta-side apartment. Contact was made with the crew soon after reaching the rear door of the apartment; they were close by and only needed to be reoriented to the egress path. The mayday was promptly terminated, and the incident commander ordered the immediate evacuation of all crews for defensive operations. As we exited, the third floor began to light up and fire was seen venting out of the windows on the delta side where the crew had just been operating. Fortunately, the close call resulted in no injuries. Nevertheless, it was a stark reminder of the importance of effective communication, accurate assessment of the fire conditions, and timely/coordinated topside ventilation and deployment of additional resources when faced with vertical extension and involvement of the cockloft.

OPERATIONAL PRINCIPLES

Ventilation must be controlled and coordinated with members advancing the initial attack line, as well as with those members involved in the search . . . options must focus around channeling the fire and its products away from the trapped occupants, as well as ahead of the advancing hoseline.

—Deputy Chief Mike Terpak,
Jersey City FD

Tactical Ventilation

Despite its recent popularity, the concept of *tactical ventilation* has been in existence for several decades now. The term was originally coined in 1987 by Paul Grimwood, crew commander with the London Fire Brigade (LFB). While serving with the LFB in the mid-1970s, Grimwood was sent to the United States on a study detachment to observe fireground operations. Riding along with FDNY's 7th Division in the South Bronx during the "War Years," in addition to many other major metropolitan fire departments, allowed Grimwood to witness firsthand the value of the American style of ventilation operations. After returning home and reflecting on his experiences, Grimwood recognized that there could be a tremendous benefit in a fusion of the two strategies.[1]

As with most of the United Kingdom and Europe, the LFB employed an "anti-ventilation" approach utilizing "zoning" tactics (i.e., door control) in an effort to confine the fire and isolate it from the rest of the structure. Ventilation was strictly withheld until the fire had been knocked down, "unless the incident commander has identified a viable objective or reason to create openings." While its application as a fire control measure can be effective for our colleagues abroad—afforded largely by the increased resistance of their unique building construction and the subsequent stability of the compartments—it can also be detrimental to the tenability in some cases, particularly victim survivability.[2]

During his time with the LFB, Grimwood recalled several occasions where the tactic of anti-ventilation stalled or slowed the advance of firefighters. In these cases, the lack of ventilation prevented victims from receiving any relief from the buildup of smoke and fire gases in some locations, such as the top of stairwells in multiple dwellings, resulting in operational failures and even fatalities. Grimwood acknowledged that while confining the fire is certainly a priority, particularly in the early stages of the incident while handlines are being stretched, it is of equal importance to create openings to ventilate the structure to improve tenability and gain *tactical* advantage. *Tactical ventilation*, as a result, synthesizes both *isolation tactics*—occluding openings to confine the fire—and *coordinated ventilation tactics*—opening up to facilitate firefighting operations and enhance victim survivability—applied with a "clear purpose and a disciplined approach."[3] The term has been adopted by the FDNY and was included with the aforementioned revisions of their operational doctrine in 2013. The ventilation chapter of the FDNY's *Firefighting Procedures* manual includes the three supporting principles of tactical ventilation: communication, coordination, and control.[4]

Communication

On the fireground, information comes at a premium. Because every situation we face is inherently unique, intelligence gathering is always a top priority. From the moment an alarm is received, we are inundated with information in varying degrees of detail and accuracy. That information must be immediately processed utilizing our knowledge base and intuition and then translated into an appropriate action plan. In the absence of a one-size-fits-all approach, conditions will always dictate action. With so much at stake, it is incumbent on us not only to collect as much information as possible *without inhibiting prompt action* but also to disseminate the pertinent information to the other members operating at the incident. Given the infinite number of potential variables and the uncertainty of a dynamic and hostile environment, we are faced with the challenge of "attempting to make perfect decisions with imperfect information."

To minimize that differential, we must engage our senses and maintain a heightened state of awareness. By staying acutely cognizant of our surroundings, we can identify critical pieces of information as they present themselves. On arrival, this physical process begins with the initial on-scene size-up and an exterior survey. We must keenly observe the building as we approach. If we pull to the far corner of the building or past it, we are granted a visual of three sides. Gaining a full 360-degree view as soon as possible will provide a complete exterior picture. This may not be feasible on arrival due to the size or configuration of the building, and it may not be accomplished until the outside vent firefighter gains access to the rear and/or the roof firefighter gets into position and they provide their respective situation reports. In regard to ventilation, attention must be focused specifically on the building (construction, openings, and layout), the presentation of the smoke and fire, and the wind condition (velocity). We must remember that size-up is not a single act. It is a continuous process that must be diligently carried out for the duration of the incident.

Not all of the information we obtain on the fireground is acquired firsthand. In fact, a great deal is received through radio transmissions (fig. 6–1). By always keeping an ear to the radio, we can ascertain the whereabouts of other crews operating within the structure, along with their assigned tasks and progress and the conditions they are encountering. This information is of particular importance to those responsible for performing ventilation. Amidst the commotion of the fireground, it is easy to succumb to the effects of stress and target fixation. We must refrain from becoming so overly task-focused that we lose sight of what is going on around us, leading to tunnel vision and auditory exclusion.[5]

Accountability is absolutely paramount when executing ventilation operations. Those who are tasked with performing ventilation, particularly when operating independently or remotely, must maintain a broader focus since their actions can single-handedly and swiftly impact the course of an incident. All ventilation must have a specific objective and support the operation, aimed toward accomplishing the overall mission. Direct communication with the interior crews (namely

Figure 6–1. The IC confers with his senior company officer to get an update on the interior conditions.
Courtesy: Paul Walsh

the officer in charge of the fire floor) and/or the incident commander is critical to obtaining the information necessary to develop an appropriate ventilation plan. Vital information must be obtained and shared, and the degree to which these tasks are carried out successfully usually determines the outcome of the operation. The need for vigilant situational awareness, coupled with clear and concise communication, is absolutely paramount.

Coordination

Coordination can be one of those words that we routinely use or acknowledge without actually understanding what it entails. We often reduce this rather complex process down to the oversimplified task of communicating with and tracking the status of the other units operating on the fireground (fig. 6–2).

While the specific manner in which it is accomplished is unique to each incident we face, the fundamental concepts should be universal. The groundwork for its success is laid well before the alarm of fire is even received. The ability to have coordination on the fireground begins at the organizational level. Departments must be well structured and well trained to attain this level of performance on a *consistent* basis. This is done by putting systems in place to predetermine and address the roles, responsibilities, and priorities on the fireground. Establishing a baseline framework of standard operating procedures/guidelines and unit/riding assignments that is then reinforced through training will maximize operational efficiency, continuity, and accountability, allowing everyone to be operating from the same playbook.

Figure 6–2. Firefighters on the roof communicate their progress and the conditions encountered.
Courtesy: Lloyd Mitchell

To ensure that tenability is maximized, firefighting efforts are facilitated, and victim survivability is enhanced, it is absolutely imperative that any openings made in the building are correctly placed, sized, and most importantly, timed. These are the three key elements of successful ventilation operations. When executed *preemptively* (prior to fire extinguishment), ventilation has a finite window of effectiveness that is dictated by the specific conditions and variables faced on the fireground. As our working environment continues to evolve, fires are reacting to ventilation at a much more rapid pace, diminishing that grace period and our margin for error. Because this condensed timeline provides the fire with an even greater advantage, it places more emphasis on timing to properly coordinate ventilation operations.

Once that window of effectiveness expires, conditions will begin to deteriorate, which can eventually lead to a ventilation-induced flashover. In essence, if air is added to the fire and water is not applied in the appropriate time frame, the fire gets larger and the hazards to firefighters and victims increase.[6] Ventilation, therefore, should not occur until a charged handline is advancing in and is able to reach the seat of the fire, unless the affected area is or can be isolated from the fire or in order to initiate the rescue of a *confirmed* victim. The success of ventilation is thus contingent on the ability to coordinate its execution with the progress of the interior crews. Aside from direct communication, doing so is achieved by monitoring the radio traffic for benchmarks and continuously assessing the conditions for cues indicating the progress (fig. 6–3).

Figure 6–3. Cuts are extended down the roof to expand the hole as it becomes consumed by the fire.
Courtesy: Curt Isakson

Control

The decision to ventilate must be based primarily on the life hazard and the progress of the interior crews. Each incident we face will present its own unique set of variables that must be specifically accounted for and addressed accordingly, especially in regard to ventilation. The size-up process must therefore include a more detailed assessment of the conditions. The air flow within the building and the ability to manage it must be evaluated to determine the appropriate interventions. We must understand how controlling existing openings and creating additional ones will affect the fire, our operations, and most importantly, potential victims. By identifying the intake and exhaust pathways, the appropriate interventions can be more accurately selected and properly executed. Doing so will also aid in the critical task of assessing the fire's rate of growth and its direction of travel, as fire burns proportionately to the quantity and source of air that it receives.[7]

Ventilation has always been predominantly focused on the exhaust component, when in reality equal attention must be placed on the intake. Because today's fires are so overly fuel rich, their development and spread are almost exclusively dictated by the availability of oxygen. Any opening in the building must be viewed as a source of ventilation and a means for feeding air to the underventilated fire. If there is a possibility that openings might intensify the fire, they should not be created until charged lines are in position, *except in matters of life safety.*[8] We must be conscious of the fact that whenever we access the building at any point, we are ventilating it. In fact, a doorway creates one of the largest horizontal ventilation openings, as it spans nearly the entire height of the wall. Doorways typically allow fresh air in and hot gases out, producing a bidirectional flow and possessing an average heat release rate capacity of approximately 2 MW (1,900 Btu/sec), the equivalent of two average-sized windows.[9] Simply leaving a door open before a handline is charged and ready to make entry can be the catalyst for rapidly reenergizing an underventilated fire.

By temporarily restricting the supply of air (without inhibiting vital operations) until a charged handline is advancing in, the fire can be kept in an underdeveloped state, promoting decay until you are in an advantageous position.[10] The mere act of controlling doors drastically reduces the air supply available to the fire and confines the fire and isolates remote areas. By limiting the fire's growth and spread, as well as reducing the migration of toxic gases and slowing the descent of the smoke layer, tenability is enhanced, increasing the degree of survivable space.[11]

The fire service has been preaching this procedure for decades when performing targeted searches (vent-enter-search [VES]), so much so that the letter

I (for *isolate*) has been recently incorporated into the acronym, making it *VEIS*. Yet this has not always been the case for conventional search operations. By closing the door upon entry, the room becomes isolated from the area of involvement. In 2000, a veteran ladder company lieutenant from the Bronx was quoted on the use of door control during the primary search at apartment fires, stating, "The calming effect on a growing fire that results from the simple act of closing the door behind you can be quite astonishing."[12] Doing so restricts further contamination and offers protection from the fire,

> Ventilation is fundamentally a matter of control, both of the air flow within the structure *and of ourselves.*

creating an area of refuge.[13] The same principle can also be applied to ventilation of enclosed stairwells, particularly in multiple dwellings. As long as the fire floor door and/or the fire apartment door is controlled, effectively isolating the stairwell from the fire, in most cases it can be ventilated, even before the onset of extinguishment.[14] In the event a potential wind-impacted condition exists, however, venting the bulkhead and/or the skylight should be delayed until the fire is put in check unless assurance is given that the access to the stairwell can remain controlled until that time.

Ventilation is fundamentally a matter of control, both of the air flow within the structure *and of ourselves.* The actions we take will dictate the fire's progression, the survivability of victims, and the well-being of firefighters (fig. 6–4).

Ventilation can make or break the outcome of an incident. Successful execution requires strict fireground discipline and is achieved by adhering to the principles of tactical ventilation (communication, coordination, and control), as mentioned previously.[15]

Figure 6–4. A firefighter displays positional discipline as he waits to vent the front windows.
Courtesy: Phil Cohen

74 *Coordinating Ventilation*

Positive Reinforcement

On a weekday afternoon during the summer, my company (led by my opposite man, Lieutenant Chris Belanger) was dispatched to a reported structure fire. The fire was in Engine 2's district, but because of scheduled training, they were out of service. Ladder 2 was the first to arrive on scene and reported fire venting out from two windows in the alpha/delta corner on the top floor of a three-story, wood-frame, multiple-dwelling apartment building. Engine 1 arrived shortly thereafter and began stretching a 1¾" handline. Knowing the building, the officer of Ladder 2 tipped off Engine 1 as to the best point of entry and stairwell to use to access the fire. He then communicated that he would be moving in ahead to search the apartment with the "irons" firefighter.

The Ladder 2 inside crew entered the fire apartment and found the front bedroom to be fully involved (fig. 6–5). Because the self-vented fire had fully consumed both windows—exceeding their exhaust capacity—it was now starting to migrate out of the room, the doorway being the next available point of low pressure. With the fire spreading and conditions deteriorating, the crew quickly traversed the living room and shut the bedroom door. By controlling the doorway, they were able to effectively confine the fire to the room of origin. This critical action not

Figure 6–5. Street view
Courtesy: Lauren Burns

only restricted the fire, keeping it in check while the handline was being advanced, but it also preserved the tenability within the rest of the apartment, allowing firefighters to conduct the primary search.

Moments later, Engine 1 arrived with the handline and positioned outside of the bedroom. As soon as the door was opened, the doorway became the path of least resistance, and the fire blew out of the room, literally skimming over the top of the nozzle firefighter's helmet. Because the energy output of the fire had overwhelmed the ventilation potential of the two windows, pressure had built up inside the room and was suddenly released when the Engine 1 crew made entry. They immediately began flowing water to knock down the flames overhead. With the stream being worked around, entraining air from behind the line, as well as cooling the fuels and contracting the gases, the fire and its by-products were driven back. Since the fire was horizontally vented directly opposite their approach, they were able to push right in and rapidly extinguish the fire (fig. 6–6).

Due to the effective communication between the engine and ladder company, as well as their aggressive, well-coordinated operations, the fire was promptly brought under control. Thankfully, the searches proved negative and property conservation was maximized. Had any victims been trapped, however, the immediate occupation of the interior and the subsequent isolation measures taken by

Figure 6–6. Signs of extinguishment
Courtesy: Lauren Burns

the first-arriving company, Ladder 2, would have resulted in their immediate extraction and the best chance of survival. These actions not only enhanced the tenability within the apartment but also facilitated the fire attack. Because of the extinguishment efforts by the Engine 1 crew, the fire was rapidly knocked down with no extension. By embodying the three operational practices—communication, coordination, and control—the primary fireground objectives were supported and the overall mission was achieved.

OPERATIONAL PRACTICES

Whoever controls the supply of oxygen wins the battle. For if it is allowed to fall into the hands of the enemy at the wrong place, it will use it to grow stronger, possibly with explosive speed. The commanders who master the art of warfare known as ventilation have a great force on their side.

—Deputy Assistant Chief John Norman, FDNY

Guideline 1
Communication

Communicate with the officer in charge of the fire floor prior to executing ventilation operations

Before ventilation operations are executed, there should be direct communication with the interior crews, especially those operating on the fire floor. These transmissions will be made predominantly through the engine company officer, who is typically in the best position to gauge the progress of the interior operations and will likely be on the receiving end of the ventilation. Firefighters tasked with performing ventilation must closely monitor the radio traffic for situation reports, obtaining the essential information to successfully execute their operations. Ideally following the *CAN* format, these updates provide the *conditions* being encountered, the *actions* being taken, and the immediate *needs* of that particular unit. It is imperative that the venting firefighters ascertain the location and extent of the fire, any control measures that are in place, the progress of the handline, and the status of the search and rescue effort (fig. 7–1).

In regard to venting for extinguishment, the request for ventilation is often initiated by the engine company officer, announcing that they are "moving in" on the fire. If staffing is limited, however, they may be actively engaged in the advance, which could hinder them from making that transmission at the ideal moment. As a result, the initial radio call may be made by the outside vent or roof firefighter, who announces that he or she is "in position" and identifies the location and the intended vent point. This prompt will get the engine company officer's attention and elicit the confirmation ("Vent, vent, vent") or postponement ("Hold off") of the ventilation. Note that when giving the order to delay or stop ventilation, it is preferable to say, "Hold off" rather than "Don't take the windows" or "Don't take the roof." This is because the first part of the message—the word "don't"—may not be transmitted or received clearly over the radio, potentially resulting in the undesired outcome.[1] If no response is provided, however, the venting firefighter must rely on conclusive visual and/or audible cues to coordinate the venting operation. Once assured that the engine company is extinguishing the fire, the venting firefighters can make openings opposite and/or above their advance.

While venting for search, specifically above enclosed stairwells, it is critical that the officer in charge of the fire floor be notified before creating any openings, as the venting firefighters will be operating remotely on the roof. This allows the venting firefighters to verify the progress of the fire attack and receive updates on the status of door control as well. If the engine company is not yet moving in on the fire, the door to the fire floor and/or the fire apartment must be controlled prior to the stairwell being vented, ensuring it remains isolated from the fire until suppression efforts are effectively underway. The venting firefighters must identify and immediately communicate the presence of a wind condition, as this can

Figure 7–1. The IC communicates with the crews operating.
Courtesy: Paul Walsh

greatly impact all operations on the fireground, particularly the aforementioned ventilation tactic. This is a major concern at fires in high-rise buildings, as the presence and severity of the wind becomes more prominent at higher elevations.

Communicating prior to venting for search also provides an essential layer of accountability. This is of particular importance when firefighters are venting a window to access a room for the purpose of performing a targeted search (i.e., VES), ensuring their exact location is made known to the incident commander and to the engine company making the attack. The searching firefighter must immediately make notification if the room cannot be isolated or if it will be delayed, as conditions will be further impacted if the fire is not yet in check. These communications also ensure that the engine company remains in position to protect the means of egress until the searching firefighters are able to withdraw from that area and/or exit the building.

All radio transmissions should adhere to the three *C*'s of communication: calm, clear, and concise.

> **Calm.** An important tenet to remember is that attitude is infectious. The fireground is stressful enough as it is, so be sure not to add to the stress by losing your composure. A good practice to incorporate is to briefly pause and take a deep breath prior to transmitting. This action serves as a mental "reboot" and allows you a moment to collect your thoughts and recenter your emotions. It also allows for the momentary lag time that many radios experience after the mic is keyed before a message can be successfully delivered. A momentary pause will prevent the beginning of the transmission from being cut off, which could otherwise prove detrimental to any subsequent operations as previously described.
>
> Projecting your voice in a strong, yet even-keeled manner sets a positive tone for all those who hear it, instilling confidence and promoting similar behavior in others. Staying calm also facilitates accurate transmission of the message to those on the receiving end.
>
> **Clear.** The more control someone has over his or her voice, limiting inflections and enunciating properly, the easier it will be to comprehend that person. Clarity of speech is vital since communication may already be difficult through the SCBA facepiece and due to the ambient noise of the fireground. *Clarity* in this case also refers to the more pressing matter of the intent of the message itself. The receiver must *fully* understand what it is the sender is stating or requesting in the message.

Concise. Effective communication must be done in the most efficient manner possible. Concisely getting your point across limits unnecessary radio traffic and ensures that the information is disseminated rapidly, which is often one of the deciding factors in the overall success of the incident.

Guideline 2
Door Control

Control the door to the fire area until a charged handline is in position and ready to make entry

An open door may seem harmless, but when it has a direct pathway to the fire, it will serve as a point of ventilation. This opening can provide an underventilated fire with enough oxygen to rapidly grow and spread toward that newly formed low-pressure opening. In fact, a standard doorway (roughly 20 ft^2) is twice the size of a common double-hung window. Furthermore, because it spans nearly the entire height of the wall, the opening can act as both an inlet and outlet when it is on the same level with the fire, producing a bidirectional flow. If a charged handline is not in position and ready to make immediate entry, *the door must be controlled until that time.* This does not, however, preclude a ladder company from entering to carry out their assignment to locate the seat of the fire and perform the primary search. If conditions and procedures allow, they must simply control the door behind them after making entry. By using a latch strap or a spring clamp on the knob side of the door, the door can be closed—controlling the opening—while preventing it from latching shut or potentially reengaging the lock. The engine company and/or the incident commander should be notified prior to making entry, thus establishing accountability. While checking the rear of the building, the outside vent firefighter must also maintain control of the back entry door after checking for any victims and assessing the conditions.

Once the engine company is ready and able to advance to the seat of the fire, the door should be propped open (using a chock, hinge hanger, nail, or strap) to allow for an unobstructed advance and to reap the full intake benefits of the opening as they make their approach. If a spring clamp was applied by the ladder company, the engine company can simply relocate it to the hinge side of the door to prop it open. The additional flow of air will increase the exhaust efficiency of any outlets in the fire room, which can initially reduce temperatures as the hot gases are displaced and discharged. More importantly, the influx of air can improve conditions at the floor level where it is being entrained, increasing with proximity to the inlet. The experiments for the UL-FSRI coordinated fire attack study found that when an occupant package was located between the front door and the fire room, the toxic dose (rate of change) at the floor level (the 1' elevation mark) began to decrease shortly after the front door was opened. Furthermore, the onset of suppression is typically occurring fast enough that any subsequent fire growth will be minimal and will not deteriorate tenability within the intake pathway (fig. 7–2). As noted in the report, "While the fresh air provided by the open front door resulted in an eventual increase in temperatures, this ventilation action led to an improvement in the potentially incapacitating gas concentrations (increase in O_2 and a decrease in CO and CO_2) at the occupant assessment locations located close to the inlet portion of the flow path."[2]

When arriving on scene, the door to the fire area may be left open by a fleeing occupant or by a concerned bystander or police officer trying to help. The sense of urgency to close the door may be overlooked or may be misunderstood. Although it may take only 1–2 minutes to stretch and charge a handline and mask up for entry, this is still sufficient time for the initial air exchange to take place and the fire to begin to react to the newly found supply of oxygen and point of low pressure. The fire cannot tell the difference between the types of openings; an opening is simply a point of ventilation.[3] When left unchecked, the fire will grow and spread along that new exhaust pathway with the potential to deteriorate conditions if the handline is not in position to promptly attack the fire.

When a door is opened, especially in underventilated fires, there will be a dump of smoke that has built up within the enclosed area. This exhaust effect

Figure 7–2. Smoke pours out and lifts after the door is opened.
Courtesy: Chris Saraceno

can produce a lift of the smoke, creating space at the floor level and allowing fresh air to flow in below the thermal balance. That tunnel or void under the smoke can provide *temporary* relief for potential victims at the floor level and can also enhance visibility.[4] On entering, seize this opportunity to scan the entryway for victims (roughly 10% of all victims are found within 6' of an exterior door and 15% within 6' of an interior door) and determine the layout of the space and the fire's location.[5] For the best view of the interior, get low—"belly down"—and shine a hand light below the smoke. When doing so, also remember to sweep behind the door for victims who may have succumbed to the smoke and look for another door, opening, or stairway. (Note that the front door predominantly swings toward the stairs in multiple-story private dwellings.) This preemptive scan should also be completed prior to controlling a door that was left open, as previously described.

As with any operation, Murphy's Law can come into play at any moment. If the fire is well advanced, it may be blowing out of an open doorway or it may have burned through the top of a closed door. Any company operating without a handline should have a member carrying a 2.5-gallon pressurized water extinguisher. Also known as "the can," it can be used to knock back a considerable amount of fire. Doing so may provide enough cover to allow another firefighter to move in and hook and control the door with a hand tool, preferably a 6' pike pole for greater reach. In the case of a burn-through situation or worse yet if no door is present, an adjacent door can be forcibly removed to cover the open doorway. When "jacking" another door, be sure it is an entry or room door unless time or availability prevents this, as closet doors are typically smaller and may be louvered. Once the door is removed, prop it up over the doorway to cover as much of the opening as possible, using a tool to secure it if necessary (fig. 7–3). Ensure the situation is immediately communicated so that the engine company knows the exact location of the fire and the incident commander is aware of the situation and its urgency. With the fire located and confined, the company can then begin conducting the primary search as conditions allow.

There are only a few *potential* exceptions where it may prove beneficial to *temporarily* extend the duration of door control beyond making entry with a charged handline. When this occurs, the door will likely be only partially occluded to allow the handline to pass through and minimize the interference with its advance. The first instance may occur when entering the fire floor if the exact location of the fire is not already known and it is not readily identifiable by the typical lifting/tunneling of the smoke upon opening the door which can occur if the fire is deep-seated or within an uncompartmentalized space (e.g., large open floor-plan homes or commercial occupancies). Opening the door and briefly pausing to "let it blow" will typically allow fresh air to move toward the oxygen-starved fire, thus prompting it to show itself as it "comes out to eat." If the fire is concealed by a severe smoke condition (camouflaged by its darkness and

Figure 7–3. An adjacent door was removed and placed over the entryway to isolate the exposed room.
Courtesy: Brian Olson

thickness) because the fire is located within the voids, or the open access door is blocking an entryway to a room or hallway where the fire is located, however, it may alter the expected reaction of the smoke and presentation of the fire from that on-level opening. Door control can be briefly maintained to check the immediate area, particularly behind the door and in the overhead while attempting to locate the seat of the fire by listening for the crackling of the fuels, feeling for the heat, and scanning with the thermal imaging camera. This can help prevent the nozzle team from inadvertently overcommitting past the fire's leading edge or in the wrong direction. It also can help prevent the fire from intensifying and breaking out above or behind the nozzle team (toward the open entryway). Once the seat of the fire is located, however, the door should be immediately reopened and secured in place to facilitate the extinguishment and search efforts.

Another more pressing situation may occur when the interior conditions are severely impeding the engine company's advance (i.e., high contents-loading or hoarding), the fire having the potential to outpace the forward progress of the handline. Additional door control may be needed to prevent the fire from rapidly developing and spreading toward the opening and over the top of the crew *if they are unable to effectively flow and move on the approach*. Once they are able to make the push to the seat of the fire, the door is then reopened completely and secured in place, maximizing the air exchange as they extinguish the fire. This may also be the case when the fire is located below grade in the basement or cellar and there is no direct access from the exterior, requiring the engine company to advance down the exposed interior stairs. *If necessary*, controlling the entry door (which essentially functions as a vertical vent opening behind them) can limit the exhaust potential and the fire spread up the stairs. Once the nozzle team descends the stairs and is able to effectively flow and move to the seat of the fire, the door should be reopened. The same approach can be applied when a wind-impacted condition exists and access to the fire is limited. If the open entry door is compromising the fire attack, particularly when on the downwind or exhaust side, the doorway should remain controlled *only until the threat is negated*. Another more likely application of this extended door control is to limit smoke contamination and preserve the tenability of common areas used for occupant egress, specifically enclosed stairwells and common hallways.[6]

The common denominator in the majority of these cases is that the opening serving as the point of entry is not functioning as an inlet and/or the conditions are inhibiting the handline from reaching the seat of the fire. Maintaining manual door control while efficiently advancing a charged handline through that entryway, however, can be challenging, even with a dedicated "door" firefighter assigned to manage it. An alternative option to consider in some of these instances may be deploying a smoke curtain. This device can be temporarily put in place to partially or completely occlude the opening, depending on whether its intake capability is desired.[7] Keep in mind, though, *the spread of fire and its by-products can be largely overcome when a properly positioned handline is flowing and moving on the approach*, especially when the fire is vented opposite/above. These additional measures, therefore, should be reserved for extenuating circumstances where its use is imperative for the engine company to successfully get into position to make the push or when victim survivability and egress will be further enhanced by their use.[8] Otherwise, just keep it simple: chock the door, stay low, and flow.

Guideline 3
Wind and Access

Access the building from the upwind side when a wind-impacted condition exists

The wind velocity should be an immediate consideration when arriving on scene. No other environmental condition can be as detrimental to firefighting efforts as the wind. The seemingly basic decision of selecting the point of entry can make or break the operation when a wind-impacted condition exists. Ideally, the building and fire should be accessed from the wind's flanks, where it will have the least amount of impact on the approach to the seat of the fire. If that is not a viable option, entry should be made from the *upwind side*, the side being impacted by the wind.

While having the wind at your back will increase the rate of air entrainment toward the fire, it will also drive the by-products away from the approach, providing relief *when a downwind opening is present*. When a sufficient opening opposite the fire is not available to serve as an exhaust point, one option to consider is to temporarily control the entry door until a handline is in position to effectively flow water on the fire or until sufficient ventilation is executed to provide the necessary relief. Doing so can prevent the wind, which will act like a ventilation fan, from overpressurizing the space, resulting in large-scale mixing of the environment. Depending on the velocity, the influx of air could cause the fire to grow and spread rapidly, or at the very least, it could disrupt the thermal balance and further minimize visibility and tenability for the victims.

Entering from the downwind side, on the other hand, places the crew in a dangerous position, especially when an upwind opening is present. The access doorway will become a dedicated outlet as the force of the wind envelops the opening on the opposite side, serving as a pure inlet and resulting in a unidirectional flow. Entering from the downwind side door would then require an advance through the fire's exhaust path. This area has been dramatically referred to as "the barrel of the shotgun" because it is between where the fire is and where it wants to go. Any notable wind speed can create harsh conditions on the approach, producing high convective heat and rapid fire development and spread.[9] If the only point of access is on the downwind side, the entry door can be controlled *if it is inhibiting the fire attack* until the fire is put in check.

If access is limited and conditions are extreme, alternative measures can be taken. A shared wall could be breached from an adjacent space to initially get water on the fire, knocking it back so entry can be made to the involved area. Wind is of particular concern at upper-floor fires, as it becomes more prominent at higher elevations. Extensive research and experimentation has been conducted on controlling wind-driven fires, specifically at high-rise buildings. Several notable pieces of equipment have been developed and employed to combat these challenging fires. In regard to ventilation, a wind-control device (WCD) called the "KO Fire Curtain" may be employed. This is a fire-retardant tarp that can be lowered from above the fire floor and positioned to cover the upwind opening, occluding the inlet and preventing the wind from entering the fire floor, thus negating its effect (fig. 7–4).

Smoke curtains can also be used to control the access door to the fire floor and/or the fire apartment. Doing so not only can help control those downwind openings, restricting the effects of the wind on firefighting operations, but can also limit the smoke contamination into common hallways and enclosed stairwells.

86 *Coordinating Ventilation*

Figure 7–4. Firefighters deploy the KO Fire Curtain.
Courtesy: NIST

Positive Reinforcement

Late one autumn night, we received multiple calls for a reported structure fire only a few blocks from the fire house. As we crested the slight grade in the road, the block was lit up by the fire, which was impressively blowing out of the building. A single window on the first floor, charlie side, had self-vented, and a large volume of fire was autoexposing up the two-story, wood-/balloon-frame multiple dwelling, which was sheathed in asphalt siding. With the building situated on a corner lot, we were granted a view of all four sides as we made the turn and pulled past the far side. Other than the one window venting fire, the structure was all buttoned up, and there was no sign of any vertical extension on the interior.

Although there was significant autoexposure extending up to the roofline, I knew that we had time. The soffits were solid—constructed of tongue-and-groove wood planks—and would stave off the fire from penetrating the cockloft long enough for us to get into position. As we stretched our handline the short distance to the front door, we did not pick up on the wind condition that was present. It was not sustained and only gusted at random intervals. As it was the only point of entry and situated directly opposite the vented window, we were advancing *downwind* of the fire (fig. 7–5).

While we were waiting for the line to be charged, the ladder company's inside crew made entry ahead of us in an attempt to confine the fire and begin their primary search. When the door was opened, they were greeted by a large volume of thick, dark smoke exhausting under pressure, *occupying the majority of the entryway*. The irons firefighter was the first one through the door and encountered high heat and almost zero visibility; the glow of the fire was the only thing discernable. With the fire extending out from the back bedroom (the area of origin), the kitchen/dining area they had entered was beginning to light up—rolling overhead. The officer immediately ordered their withdrawal, and the door was controlled as they exited.

Figure 7–5. The alpha/delta corner
Courtesy: Michael Carenza Jr.

Once our line was charged and flow checked, the door was reopened, and we advanced in first. We were unaware that there was a gusting headwind, so the door was chocked, and the storm door was strapped fully open, with the strap attached to the porch railing. With the entryway serving as a downwind outlet, we were now between where the fire was and where it wanted to go—"the barrel of the shotgun." As I was loading hose into the kitchen so we could reach the back bedroom, I could hear and feel my nozzle firefighter briefly opening up and then closing the nozzle, attempting to "hit and move" to advance. With the wind stoking the fire and driving it in our direction, however, conditions would rebound after he shut down the line, and it would light up again immediately. The incident commander reported after the incident was over that during this time, the wind gusts were actually overwhelming the window and preventing the fire from venting out, transforming the outlet to an inlet and producing a unidirectional flow. After the second time the nozzle firefighter completed this cycle, I yelled over to him to keep the bale wide open and that we would have to make a "push," flowing and moving in tandem all the way to the seat of the fire.

With the line operating at full bore and enough hose staged, we married up (in the "crooked-lean" position) and started our advance. As the nozzle firefighter directed the stream in an O-pattern—concentrated out front and overhead—we were able to entrain enough air to create our own pressure front and seal off the fire room. Being able to counteract the force of the wind and keep the fire at bay allowed us to continuously advance to the seat of the fire and achieve knockdown. Once we hit the back ("dry") wall and extinguished the fire room, my nozzle firefighter leaned out of the window and scrubbed the outside wall and swept the soffit to knock down the exterior fire (fig. 7–6). Thankfully, the approach path was a straight shot and only about 10' to 15', with no obstructions other than the kitchen table set. If the fire location or the interior conditions been more complicated, the wind velocity been sustained or more severe, or had we lacked the ability to continuously flow and move, we may not have been as successful with the downwind entry door uncontrolled.

Figure 7–6. The nozzle firefighter scrubs the exterior after knocking down the main body of fire.
Courtesy: NBFD

Guideline 4
Wind and Ventilation

Use the sides of the building that are not being impacted by the wind when venting for extinguishment

When selecting the location to create a ventilation opening for the advancing engine company, the direction of the wind must first be identified. A sustained wind velocity of as little as 10 mph can noticeably affect the fire's behavior and thus firefighting efforts. At 20 mph and above, the effects of the wind can become extreme, resulting in high-velocity convective heat currents and rapid fire growth and spread.[10] The upwind side of the building—the side that is being impacted—should not be used when venting to support the extinguishment effort. If an opening is created on that side, the wind will then flow in through it, causing the opening to function, at least partially, as an inlet.[11] With the wind competing for the same opening, the exhaust capacity will be restricted, with the potential to overwhelm it altogether (fig. 7–7).

The degree to which it does so will depend on the wind velocity, as well as the pitch of the roof when vertically venting. The greater the speed and the more directly the air is able to flow through the opening, the less it will function as an outlet. Not only will this stifle the by-products from being released, but the rapid influx of fresh air directly to the fire can adversely affect the fire attack. Flowing in the direction of the approaching engine company, the wind can intensify the fire and may at least partially redirect the by-products in their direction, slowing or potentially stopping their advance altogether.

The location of the fire, and the subsequent availability of windows within that area, may prevent the use of horizontal ventilation for extinguishment. If the fire is located in the attic/cockloft space or the floor directly beneath it, vertical ventilation can still be a viable option, as at least a portion of the roof will not be impacted by the wind. In the case of a peaked-roof dwelling, the downwind slope must be used for any cut operations. Assessing and communicating the wind velocity, therefore, is imperative to determining the proper location and method for opening up (fig. 7–8).

Figure 7–7. Vent the windows not affected by the wind.
Courtesy: Nicholas Papa

90 Coordinating Ventilation

Figure 7–8. Vent the downwind face of the roof when a wind-impacted condition exists.
Courtesy: Nicholas Papa

Guideline 5
Exposures

Vent in a manner that does not create an exposure problem

While this may seem obvious, there are several precautions that must be taken to ensure ventilation does not endanger any exposures, whether interior or exterior. Many of the interior exposure hazards can be addressed through the door control measures previously covered, and more importantly, by the proper placement and application of handlines. To prevent the threat of autoexposure, use caution when venting directly below open windows, especially when window treatments are exposed, or within porches/overhangs, or when combustible siding is present (fig. 7–9). If such openings are the only option, be sure the fire is in check by the attacking handline before taking them, ensuring the fire does not lap up the side of the building, autoexposing to upper floors. The same is true for windows that serve fire escapes and balconies. If they are being used for access/egress, window venting below must be withheld, as members operating above could be cut off. Also, if any building occupants are attempting to escape or seek

Figure 7–9. Open wood porches present a significant exposure hazard.
Courtesy: Matt Kelly

Figure 7–10. These wood-frame multiple dwellings, sheathed in asphalt ("gasoline") siding, are only about 10' apart.
Courtesy: Nicholas Papa

refuge, they can quickly become endangered by smoke and/or flames from below, potentially trapping and severely injuring them.[12]

The other more traditional exposure hazard consists of neighboring buildings in close proximity. This can be especially problematic when these adjacent structures are wrapped in combustible sheathing or siding, notably asphalt shingles ("gasoline siding") and wall cladding systems (such as exterior insulation and finish systems [EIFS]) (fig. 7–10). When these conditions exist, attempt to utilize another side of the fire building for window venting or hold off on taking the glass on that exposure side until the advancing engine company is effectively extinguishing the fire, as previously described.

Guideline 6
Vent for Extinguishment—Horizontal

Ensure fire attack is underway prior to horizontally venting for extinguishment

The timing aspect of horizontal ventilation for extinguishment is one of the most critical aspects to the success of the fire attack. If it comes too early, the air exchange will cause the fire to react to the additional supply of oxygen. If water is not promptly applied, the growing fire can consume the window opening. At that point the fire will spread toward the next available outlet, the open doorway, causing it to travel toward the advancing engine company.

While there is potential for conditions to briefly improve, particularly at the floor level closest to the intake side, it will begin to expire once the growing fire overwhelms any exhaust openings created. Horizontal ventilation for this purpose must not occur until the engine company is in position to successfully advance in on the fire.[13] If the ventilation comes too late, however, the attacking crew will not receive the intended relief when they need it the most, causing them to unnecessarily "take a feed" (fig. 7–11).

When breaking glass, be sure to start with an upper corner of the window, as the topmost portion is subjected to the highest heat and may be weakened. Because heat and smoke naturally rise, doing so will exhaust the greatest amount, in addition to allowing conditions to be immediately evaluated prior to clearing the rest of the window. Taking the top section of the window first prevents a large section of glass from falling down onto the venting firefighter. To maximize the exhaust efficiency, the window should be *trimmed out*, cleaning out any remaining glass shards and removing any window treatments, the sash, and other obstructions that might occlude the opening.[14]

Figure 7–11. The firefighter positioned the ground ladder to vent both windows.
Courtesy: Nathen Maronski

There is some debate as to exactly when window venting should begin. Because the grace period prior to water application is highly variable according to the conditions present, its potential benefits can be fleeting, making it extremely challenging to accurately predict the time frame. It is most prudent, therefore, to initiate ventilation along with the onset of fire attack. At the very least, the engine company officer must be certain that the nozzle team is in position with a charged handline and is capable of advancing to the seat of the fire. The nozzle team must possess overwhelmingly superior force and dominant offensive positioning. Venting opposite the handline and minimizing the time between the creation of that opening and fire suppression capitalizes on the additional air entrainment and subsequent exhaust efficiency. When properly sequenced, the thermal and toxic exposure can be reduced, limiting the hazard to the interior crews and improving victim survivability.[15]

Effectively timing this operation can be as simple as a brief radio transmission requesting/confirming ventilation. As previously stated, however, the engine company officer may be a working member of the nozzle team (due to limited staffing) and thus unable to reach the radio mic to transmit at the ideal moment. In such instances, the outside vent/roof firefighters must be monitoring the radio traffic for key benchmarks and diligently sizing up the fireground to gauge the progress and determine the correct operational tempo. By engaging their senses, they can observe changes in the conditions such as dissipation of the fire, lightening of the smoke, steam conversion, and the handline's stream intermittently exiting the building. They can also listen for the sound of the stream impacting the interior surfaces. While it is best to receive explicit confirmation from the officer in charge of the fire floor, skilled outside vent and roof firefighters are able to reliably utilize these sensory inputs or cues to effectively time the operation.[16]

Learning the Hard Way

After making lieutenant, I was assigned back to an engine company. In the swipe of a pen, I went from being a ladder company firefighter—the one providing the ventilation—to being an engine officer—the one requesting it. Being back on the receiving end of ventilation has provided me with an entirely different perspective and a greater appreciation for the tactic. One of my early fires was a top-floor fire in a three-decker (a three-story, balloon-frame multiple dwelling with one unit per floor), our "bread and butter" (fig. 7–12). We stretched dry to the third-floor landing as another company had confirmed a *small* fire in the front bedroom and controlled the apartment door. I noted a moderate smoke condition through the crack of the door (as it was not latched shut) but could not see any active flames.

Once we had the line flaked out, I grabbed the mic of my portable radio and made the request for the line to be charged. As soon as I released the transmit button, I heard the sound of breaking glass. I had not called for ventilation, and more importantly, we did not have any water yet. To make matters worse, the windows taken were in the adjacent living room. As the line was beginning to fill with water, we saw a glow immediately appear in the front bedroom. The room quickly touched off, and fire began to roll out and across the ceiling, directly toward the newly created openings. Luckily, there was no delay in the arrival of our water. The line was bled off and flow checked, the door was opened, and the stream knocked down the fire just as quickly as it had escalated.

Figure 7–12. Street view
Courtesy: Frank Papa Jr.

My nozzle firefighter, with about a year on the job, and I received an invaluable firsthand fire behavior lesson. We were able to see just how volatile under-ventilated fires can be and the impact of premature/misplaced ventilation. The outside vent firefighter also learned the importance of confirming the fire's location and the progress of the interior crews, specifically the engine company making the attack. While the call for water is an important benchmark, prompting the outside vent firefighter to be at the ready, it is *not* an appropriate cue for when to execute ventilation.

Guideline 7
Vent for Search—Horizontal

Ensure fire attack is underway or the targeted area can be isolated prior to horizontally venting for search unless conducting a rescue

Because horizontally venting for search is conducted remote from the area of involvement, it can be initiated prior to the onset of extinguishment, with *strict* stipulations. To successfully "vent as you go" while performing a conventional primary search, the door to the room being vented must first be controlled if the fire has yet to be put in check.[17] By occluding the entryway, controlling the door permits window venting within that isolated area, as there is no longer a pathway to the fire. This tactic is essentially the inverse of the VEIS tactic.[18] By exhausting the heat and smoke built up inside the room and allowing fresh air in to replace it, tenability and visibility can substantially improve, facilitating the search effort. Performing local horizontal ventilation for search efforts is critical for those areas outside of the *intake pathway* (between the front door and any other inlets and the fire) since they otherwise will remain remarkably unaffected by other sources of ventilation (i.e., venting or extinguishment).

The vented window can also serve as an alternative means of victim removal or for egress if it will further promote their survivability or the intended means is compromised. Taking the victims out the window will directly place them into fresh air. This can be preferrable to dragging them out the controlled door and subjecting them to the increased thermal and toxic threat in the contaminated hallway, especially if the fire has yet to be contained or the travel path is a considerable distance. Ideally this is accomplished when on the ground floor level, or when the window leads to a suitable porch or fire escape, or when an aerial device (ideally the bucket of a tower ladder) can be rapidly placed to the window.

While the use of ground/portable ladders is highly effective and should not be discounted or discouraged, it must be noted that taking an incapacitated and potentially burned victim out a window and down a ladder can be intensive and may require additional manpower that may not be readily available. As such, the state and size of the victim, as well as the window size and the sill height, must be considered. If those resources are not present, another option may be to temporarily shelter the victim in place until they become available or until the engine company is at least on the approach and can relieve conditions along the egress path and provide cover for the extraction. These decisions should be made on a case-by-case basis, as what is best for the victim will depend on the situation at hand (fig. 7–13).

The barrier effect of the closed door also makes the space a temporary area of refuge. In addition to limiting further contamination, even a hollow-core door can provide significant protection from fire for up to five minutes.[19] Experiments

Figure 7–13. Fire escapes and aerial devices can serve as alternative means of egress for victim removal.
Courtesy: Brian Grogan

have shown that a single door separating the fire can maintain tenable conditions for victims within an adjacent isolated space. At the lower-elevations—the floor-level where victims are likely to be present—temperatures were found to be more than 500°F lower (remaining around 100°F), and fire gas concentrations were within the survivable ranges behind the closed door (fig. 7–14).[20]

Before venting as you go, specifically when operating on the floor above, it is imperative to know the location and extent of the fire. When fire is self-venting from a window directly below and autoexposing up the side of the building, taking a window can allow the fire to extend into the room being searched. If the window has failed or starts to fail, a door (ideally other than the one serving the entryway for the room) can be removed and used to cover the window opening, while using a "can" to knock back any extending fire. Once the search of the room is completed or conditions force a retreat, the door to the room must be controlled upon exiting, isolating the rest of the floor/apartment from the fire.

In the event there is a *confirmed life hazard* (information about the last known or suspected location of the trapped victim), an exception can be made. To conduct an immediate rescue, a window can be vented (ideally from a stable platform, such as a porch roof or the bucket of a tower ladder) to initiate a targeted search (VES) of the identified room, whether or not the space is isolated or capable of being isolated or if the fire attack is even underway. Although this form of venting for search can knowingly draw smoke and fire to the opening being created, it will then provide direct access to where the victims are reportedly located. By drastically expediting the search and rescue operation, the victims will be given the best possible chance of survival.[21]

When doing so, it is vitally important to evaluate the conditions once the window has been vented, pausing for a few seconds to determine if entry is viable. If the doorway is uncontrolled (as evident by exhausting smoke), the operational

Figure 7–14. Heavy soot staining, blistering, and char are present in the fire area, while the adjacent bedroom, isolated by a closed door, is undamaged.
Courtesy: SFD

time limit must be predicted. Ask yourself, "Will I be able to search and get out in time?"[22] If possible, utilize a thermal imaging camera to read the temperatures within the space. This assessment period also provides the opportunity for the environment to react to the sudden change in air flow ("letting it blow"), keeping you out of harm's way if a rapid fire spread event results. Once committed to making entry, the conditions must be diligently monitored throughout the search-and-rescue effort. If the door is open, the fire will begin migrating to the new low-pressure area the instant the window is vented (fig. 7–15).

To maximize that operational period, there are certain precautions that must be taken. If deploying a ground/portable ladder to the window, do not use it to vent the glass. Once the window is broken, the clock starts ticking on how much time you have to control the doorway and perform the search and rescue. The increasing speed with which fire reacts to ventilation due to newer construction methods and modern fuel loading has decreased this window of opportunity. Furthermore, the use of energy-efficient (multipane/gas-infused) and hurricane-rated/impact-resistant windows can render this technique inefficient at best. Much more physical force will be required, increasing fatigue, and it may produce enough of a shock load to unlock the "dawgs" (pawls) on an extension ladder, causing the fly section to suddenly retract if the halyard is not tied off.[23]

No matter how well this technique may be performed, the window will still have to be trimmed out to remove any remaining glass shards, pieces of the sash, or window treatments that may be in the way. Such an approach will be further hindered when a window-mounted air conditioning unit is present. The preferred method for performing this operation is to set the ladder just beneath the

Figure 7–15. A firefighter "hurdles" over the sill to make entry to search.
Courtesy: BC2 Fred Ruff, #JobTown

windowsill and at a lower angle, which is ideal for access/egress and rescue. The searching firefighter should then ascend the ladder, keeping his or her head level with the sill to provide cover in the event of an explosive/flaming discharge. The searching firefighter should then go on air and use a long hand tool (preferably a 6' pike pole) to break the glass. This puts the firefighter in an action position, ready to make immediate entry (fig. 7–16). Prior to traversing the window, the floor should be swept and sounded, and then the room must be promptly isolated.

A reliable rule of thumb for locating the interior doorway on entering a room from the exterior, called "the corner trick," is to move diagonally in direction from the outside corner of the building[24] (fig. 7–17). For center rooms, simply cut across the room, and the doorway should be located along that opposite wall (typically

Figure 7–16. The firefighter is in position to vent the window and make entry.
Courtesy: Jim Moss

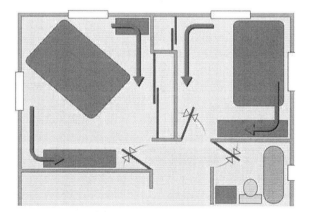

Figure 7–17. The "corner trick" displays how traveling away from the outside corner of the building can lead you to the interior door.
Source: Frank Ricci and Josh Miller, "Aggressive and Practical Search," *Fire Engineering* (March 1, 2016).

toward one of the two sides). Where only one window is present within the room, the door should be straight ahead.[25] Isolating the space should remain the top priority, particularly when the fire is not in check. Even if a victim is located while traversing the room, it is critical that firefighters press on to control the doorway.[26] Isolating the space will extend the operational window and can be the ultimate deciding factor concerning whether or not that rescue can be successfully executed. As the saying goes, "Take time to make time."

Whether or not the interior door is open, get a quick visual on the hallway since any victims attempting to escape may have collapsed just outside the targeted room. Doing so will also aid in assessing the fire conditions and your operational limitations. On completion, immediately control the door to reisolate the room being searched. When closing the door (if present), take note of the construction, gauging the material and the weight to determine how well it will resist fire.[27] If the doorway can be adequately controlled and conditions allow, the search can be continued beyond the targeted room if the initial search turns up negative, after ensuring that the incident commander and the engine company are first made aware for accountability. To provide a secondary means of egress, the rapid intervention team or an on-deck crew can deploy additional ground ladders to the areas being searched next. If the door is missing or cannot be shut due to obstructions, or it is otherwise not possible to control the opening, the closest, most suitable replacement should be used to isolate the room. Ideally, this is done by removing a neighboring door (preferably not from a closet) and propping it up in the doorway.[28] Greater caution should be exercised in this case since the opening may not be fully occluded and the fire will react accordingly.

Vent-enter-(isolate)-search is highly effective and has proven its worth many times over. While it does carry an inherent risk, as does every aspect of firefighting, the last known line-of-duty death attributed to the tactic occurred more than 75 years ago, when Frederick Davies of the London Fire Brigade made the ultimate sacrifice on August 22, 1945.[29] With a greater understanding of fire behavior, along with the recent public education/fire prevention campaign, "Close before You Doze," the application and success of VES should increase in frequency (fig. 7–18).

With more of the public keeping their bedroom doors closed at night, the principal threat of this operation will be essentially mitigated, as the room being searched is already isolated from the fire. This is especially important since the bedroom is by far the most common location where victims are found, and nighttime is the most likely time of entrapment. According to the latest annual findings of the Firefighter Rescue Survey, extrapolated from the civilian rescues reported in the year 2020, VES/VEIS represented the second-most utilized search tactic (executed in 60 of the 237 cases, or 25%). The most telling data point, however, is that VES/VEIS yielded the highest survival rate—a staggering 72% (43 of

Figure 7–18. The firefighter "dives in" to search.
Courtesy: Lloyd Mitchell

the 60 victims removed).[30] If we are telling the public to close their doors, we must be ready and willing to aggressively rescue them in the manner that will have the greatest impact on their survival (fig. 7–19).

Positive Reinforcement

Once when I was a private, we received multiple calls in the late evening hours on a spring weekend for a reported structure fire. On arrival, we found a small, single-story, wood-frame, bungalow-style private dwelling, with heavy fire showing from the charlie/delta corner. As I stepped off the jump seat of the engine, a neighbor greeted me and frantically communicated that the owner of the home was still inside. I immediately grabbed the irons, informed my officer, and got authorization to start searching ahead, as we had a jump on the ladder company, and they were still a few blocks out.

Figure 7–19. The outside team of FDNY's Tower Ladder 1 performs a daring VES operation under heavy fire conditions, successfully rescuing a trapped victim.
Courtesy: Lloyd Mitchell

While the handline was being stretched, I proceeded to the front door. On my way there, another neighbor yelled over and informed me that they had just seen the man inside, pointing to the second window in from the front, on the bravo side. I then relocated to that position and updated my officer of the situation. Using the halligan bar, I took the glass, prompting a dump of thick, dark smoke under pressure. Diving in, I encountered conditions that were nearly "lights out," with only a few inches of poor visibility down at the floor level. I could hear the fire crackling around the corner in the back bedroom. I quickly got my bearings and realized I was in a small dining room and had no way of isolating my position, which increased the danger and severely limited my operating time. As I made the final turn inside the room, the search proving negative, I heard a transmission come over the radio, stating that the victim had been located.

After receiving that update and sensing that the conditions were starting to deteriorate, I made my way to the front door to link up with my officer and get on the handline. As the line was being charged, the kitchen and dining room flashed over in rapid succession due to the window I had vented for access, with fire rolling across the entryway ceiling toward the now-open front door.

This event illustrates the limited time frame and potential for rapid fire spread when executing VES in an uncompartmentalized space or any room that cannot be isolated (decreasing with proximity to the fire, as well as with its severity). As such, the tactic, *in this instance*, should be reserved in such instances for confirmed entrapment with the last-known location of the victim. Once operational, we pushed in, knocking down fire in the overhead, and made it to the dining room. Over half of the living area was involved, and the fire had extended throughout the attic space. An intense attack ensued, requiring a second handline operating in tandem to bring it under control.

The victim, delirious from CO poisoning, had moved away from the window at which he had been seen and had attempted to exit via the back door (his normal means of egress), unknowingly traveling deeper into the building and past the fire. His actions reinforce the fact that victims can act irrationally and may not remain in their last-known location. Simultaneous search operations (both conventional and VES) that may require going beyond the targeted room are therefore essential to expedite the location and removal of any trapped victims.

Fortunately, the ladder company's outside vent firefighter was able to make the grab successfully. While performing his check of the rear, he gained access to the back door to recon and search the immediate area. He instantly saw the incapacitated victim within the threshold of the door, a heads-up move that paid dividends. He dragged the man out, and with the help of the ladder driver, brought him to the driveway, where he was turned over to the paramedics. The victim was rushed to the nearby hospital and transferred to the burn center soon after. Because of the aggressive search efforts, along with rapid advanced life support and transport to definitive care, the man recovered with *no deficits*.

Guideline 8
Vertical Ventilation

Ensure fire attack is underway or the targeted area is isolated prior to vertically venting unless there is potential for a backdraft or cockloft explosion on accessing the fire

Vertical ventilation can come in many forms and *must* be prioritized based on the needs and the resources available. Before committing the manpower and time, which can be intensive, ensure its intended effect is capable of being achieved, while simultaneously and efficiently completing the essential operations of fire attack and search. As with any other form of ventilation, the objective must be defined, ensuring the operation supports either of those aforementioned tasks.

When venting for extinguishment, the roof crew must confirm the location of the seat of the fire and the progress of the advancing engine company below. This form of vertical ventilation is best suited for fires in the attic/cockloft area, where the opening can be made directly above the fire's level and at the highest point (fig. 7–20). Additionally, the reduced fuel load that can be encountered (comprised only of the wooden structural members and decking) can increase the confinement ability of the vent opening created. This is a tremendous benefit, as most attic/cockloft spaces are not compartmentalized, and lateral fire spread is a chief concern (fig. 7–21).

104 *Coordinating Ventilation*

Figure 7–20. Smoke pushes out from the seams of the roof and windows of the finished/occupied attic space.
Courtesy: Giovanni Sanzo

Figure 7–21. The smoke lifts as it is channeled out of the vent hole cut into the roof, relieving conditions below.
Courtesy: Giovanni Sanzo

As with most things on the fireground, this benefit is not a guarantee. Steep roof pitches are common in both private dwellings and older multiple dwellings, particularly in regions prone to snowfall. The walkable attic space that is created can be used as an additional living space or for storage. Especially under these circumstances, well-timed vertical ventilation can provide the lift needed to effectively operate, redirecting the buildup of heat and smoke upward and outward, as well as relieving the pressure.[31] With large plastic totes being one of the most popular means of storing items (usually holiday decorations, clothes, and toys), the fuel load can easily match or even exceed that of a living space, requiring that they be addressed accordingly.

While vertical ventilation for extinguishment can also be effective for fires involving the floor beneath the attic/cockloft area, it may not be needed because suitable horizontal openings (windows) are frequently available, and the space is largely compartmentalized. There are several negative consequences that could result from cutting open the roof and breaching the ceiling below in this case. First, there will likely be considerably more property damage than from window venting. There may also be reduced exhaust capacity if the opening created in the ceiling does not match the size of the hole cut in the roof. Furthermore, there is the inherent risk of spreading fire into the previously uninvolved attic/cockloft space if the opening created is not properly timed or the fire attack effort is hindered. Finally, there is the pressing issue of resource and time constraints. If manpower is limited, unnecessarily committing additional personnel to open up the roof *when a comparable outcome can be achieved through window venting* is a misuse of resources. Adequately staffing the interior crews is paramount to ensuring the first handline can reach the seat of the fire and the primary search can be completed in the most effective and efficient manner, accomplishing the priority objectives.

Although topside ventilation creates a dedicated outlet, which maximizes the exhaust capacity and lift of the thermal balance along the intake pathway, a disciplined approach must still be taken. Even when the fire is in the attic/cockloft space, which may present a reduced fuel load, the duration of its window of effectiveness remains highly variable and can expire in short order. This does not, however, prevent the roof crew from starting if a cut operation is needed since it can take a considerable amount of time and effort. When doing so, they must be diligent to maintain the integrity of the section being cut and not louver it until the appropriate time. The only exception to this is when there is a threat of backdraft or smoke (cockloft) explosion on accessing the building/area of involvement. In this case, priority should be given to topside ventilation to ensure any explosive pressure wave (*deflagration*) produced is directed upward and outward, away from the crews standing by. The engine company must be staged on the flanks, protected by a wall or in an area of refuge with a charged handline, ready to make immediate entry/access and apply water to the seat of the fire.

Vertically venting for search, typically conducted at flat-roof multiple dwellings, is done for the purpose of relieving contaminated stairwells. Heat and smoke can migrate into the stairwells, collecting at the top floor and banking down.

This can prove deadly for occupants of upper floors who are attempting to escape, as lethal concentrations of fire gases can be readily achieved. It also reduces visibility and can slow down the advancing crews and require them to go on air—cutting into their supply—well before they reach the area of involvement, increasing fatigue and limiting their operating time. This form of vertical ventilation can be conducted rapidly, as it utilizes the natural openings of the building (fig. 7–22).

Above the stairwells in these occupancies, there is typically a bulkhead door providing access to the roof, and there may also be a skylight (fig. 7–23). As soon as confirmation is received that the door to the fire floor and/or the fire apartment is being controlled or the engine company is effectively moving in on the fire, the stairwell bulkhead and skylight can be opened fully. If a potential wind-impacted condition exists (e.g., a fire in a high-rise building), however, the creation of any downwind (vertical) openings, particularly over the fire attack stairwell, should be withheld until one or more of the following applies:

- The nozzle team has the fire in check.
- A wind control device such as the KO Fire Curtain can be deployed to occlude the upwind (intake) opening, negating the effects of the wind on the interior conditions.
- The stairwell can remain isolated, either by keeping the door shut or by utilizing a smoke curtain to control the opening.

Otherwise, these vertical points of low pressure can promote rapid fire growth and spread, significantly deteriorating conditions along that pathway.

If there is only a skylight, a *draft stop* (a secondary skylight flush with the top-floor ceiling designed to keep the conditioned air out of the skylight well) may be present below and must also be cleared out.[32] Prior to venting skylights, be sure to notify the interior crews, ideally over the radio, that the glass is being taken. An additional measure that should be taken is to rap loudly on the frame

Figure 7–22. A draft stop beneath the skylight
Courtesy: Danny Troxell

of the skylight with a pike pole and then use it to take a single panel or small section of glass in one of the corners, waiting a few seconds before cleaning out the remainder to give members below a chance to seek refuge. Be sure not to breach the *returns* (the walls framing the skylight well below) if the skylight is remote from the fire or if a handline is not in position on the top floor, as it will vent the cockloft, accelerating fire growth and drawing it to that point of low pressure.[33]

Roof firefighters serve as the eyes and ears of both the incident commander and the crews operating on the interior due to their unique vantage point. They are in one of the best positions to identify vertical fire extension and the overall configuration of the building. When arriving on the roof, they must be proactive and immediately get to work to recon the area and its perimeter, specifically in large multiple dwellings and commercial occupancies. The back side and any shaftways must be checked for smoke/fire and victims, especially in larger buildings, as these remote areas may not be readily visible or accessible on arrival (fig. 7-24).[34]

Figure 7–23. A large bulkhead with windows
Courtesy: Danny Troxell

Figure 7–24. A light shaft in the center of the building
Courtesy: Danny Troxell

While traversing the roof, immediately identify a secondary means of egress, as well as any significant loads and fire stopping in place. Examine natural openings and vents encountered for indications of fire spread within the void spaces and into the cockloft, specifically in the utility/pipe chases and dumbwaiter shaft (fig. 7–25). This may be evident by smoke or embers exhausting from vents, ducts, or shafts, pipe stacks that are hot to the touch, or bubbling tar (typically around these). Signs or reports of extension/cockloft involvement should prompt a diagnostic cut in the suspected area for confirmation and to determine its extent, *immediately* communicating the findings. The inspection cut or hole will also identify the roof construction, including the type and thickness of the decking and the structural members, along with their location. Sometimes conditions are discovered that can pose a threat to operations or will delay or prevent the vent hole from being cut, such as a built-up roof (multiple layers of roofing materials), fortified roof (steel plating), or "rain roof." In these cases, the interior crews and the incident commander must be promptly alerted so measures can be taken to adjust accordingly.

Another method of checking for extension is by forcing open a nearby scuttle hatch and breaching the returns. If fire is present, the hatch can be closed until a handline is in position to attack the cockloft (figs. 7–26 and 7–27). An additional preemptive operation that must be given priority is forcing the bulkhead door and searching the top-floor landing for potential victims. If the fire floor and/or the fire apartment door is not controlled or the fire attack is not yet underway, however, the bulkhead door must be controlled while doing so.[35] The following hierarchy can be used to prioritize the creation of openings when operating on a flat roof: bulkhead doors and stairwell skylights → room skylights and windows (for top-floor fires) → scuttles; shaftway doors/skylights; and ventilators and duct

Figure 7–25. Vent ("soil") pipes and a skylight
Courtesy: Nicholas Papa

Figure 7–26. Scuttle hatch
Courtesy: Danny Troxell

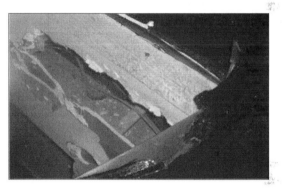

Figure 7–27. Return walls
Courtesy: Danny Troxell

vents → return walls. The sequence for peaked-roof operations is as follows: gable windows and louvers/attic vents → skylights. In either case, cutting the roof and punching down the ceiling should be the final offensive tactic. When positioning to cut, the vent hole ideally should be placed to capture the fire room and the hallway/adjoining room as well, about 8' to 10' from the roof edge on a flat roof. Orientation can be improved by looking over the side of the roof and lining up with the windows.[36]

In the event a fire in the cockloft of a flat-roof structure has overwhelmed the primary ventilation hole, and further expanding the cut will not suffice in its confinement despite the coordinated fire attack efforts occurring below, the roof operations may need to transition to a defensive posture. For large buildings with a "wing-type" design (e.g., multiple dwellings resembling the shape of the letter L, U, H, or E), or for buildings that are long and narrow, a section may have to be sacrificed to stop the lateral extension from taking over the remainder, strategically "exchanging ground for time" in order to maneuver and gain tactical advantage.[37] A *trench cut*—a 3' wide hole that spans from one edge of the roof to the opposite end—can be made as a last-ditch effort to halt the fire's progression. By cutting out the roofing materials and dropping the ceiling below, thus removing the available fuel source, it essentially functions as a structural "firebreak."[38] To facilitate this operation, perpendicular cuts should be made roughly every 4' to divide the trench into manageable sections. If additional manpower is available, the ceiling can also be pulled by crews operating on the top floor.[39]

For a trench cut to be effective, however, its outer boundaries—the building's exterior walls—should be noncombustible to prevent the fire from going around it, making it best suited for type III ordinary construction, where the wood-framed roof is supported by the brick/masonry load-bearing walls.[40] Due to the labor- and time-intensive nature of the operation, a trench cut should be placed at least 20' from the primary vent hole over the fire. To enhance the reflex time of execution, take advantage of the building's layout and features whenever possible. By selecting the narrowest point of the roof, such as the area where the wings connect (also known as the *throat*), especially where fixed structures and/or natural breaks or openings exist (e.g., shaftways, stairwells, skylights, and scuttles), the length of the trench cut will be reduced. Doing so also decreases the volume of the fire front by funneling it down through that area, increasing the suppression capabilities of the handlines when positioned at that point.[41]

This is akin to the military maxim of leveraging terrain against an opposing force. When outnumbered, selecting a natural chokepoint minimizes the magnitude of the assault. This is the same principle behind the strategy employed by the vastly outnumbered Spartans, who selected a narrow mountain pass in which to fight in the historic battle against the Persians at Thermopylae, as dramatized in the movie *300*.

While a trench cut can and should be initiated proactively, due to its time-sensitive nature it must not be pulled until or unless it is actually needed (fig. 7–28). Prematurely opening the trench will counterproductively accelerate the fire's horizontal movement by drawing it to that point of low pressure. To provide an indicator as to the correct timing, *observation holes*—triangular inspection cuts—must be placed between the trench and the vent hole. Once the fire reaches those openings and begins to show, the trench should be opened. Those inspection holes must, however, be placed far enough away (approximately 5') to give the appropriate lead time to fully open up the trench before the fire's arrival.[42]

Before this process is initiated, no one operating on the roof should be on the fire side of the trench. As a precaution, observation holes must also be placed on the opposite (clean) side to identify if the fire is overrunning the trench. A handline also needs to be in place topside to suppress any fire involving the roofing materials, as well as directly below to prevent the fire from jumping the trench.[43] It must be reiterated that the trench cut does not replace, nor should it precede, the cutting and expanding of the ventilation hole over the main body of fire unless

7 • Operational Practices   111

Figure 7–28. A view from the "blind," open shaftway of the throat of a U-shaped multiple dwelling
Courtesy: Nicholas Papa

there is a surplus of personnel and doing so does not delay the primary vent hole from being cut. In most cases the initial vent operation, when positioned, timed, and sized correctly, will provide the interior crews with the needed relief to halt the fire's progression.

Guideline 9
Void Spaces

Ensure a charged handline is in position prior to opening up walls and ceilings when accessing the attic or void spaces

Any opening that leads to the area of involvement must be viewed as a source of ventilation, even openings created in the walls and ceilings. For this reason, "opening up" (overhaul) must be treated like any other form of ventilation. If concealed fire is suspected in the void spaces, it should be isolated by delaying the creation of any openings into those areas until a charged handline is in position. Once an underventilated fire is given a fresh supply of oxygen, it will gain momentum, potentially at an overwhelming pace. Having a handline at the ready prior to opening up will ensure that any fire encountered will be put in check. In an emergency, the 2.5-gallon pressurized water extinguisher can be utilized. In the hands of a skilled operator, the "can" is a tremendous asset, capable of knocking back a considerable amount of fire. Another precaution to take when opening up ceilings to check for fire extension is to make your initial inspection hole from a doorway, ideally to a closet. In the event the fire reacts violently, such as a cockloft explosion, the doorframe will provide protection from any ceiling collapse. This is a significant hazard with plaster and lath construction, where pieces can be of significant weight. Also, if the fire begins rapidly spreading outward, the door can be immediately controlled while the handline is moved into position, effectively confining the fire to that area. To minimize the chance of this occurring, the initial inspection hole should be made as small as possible. This is best accomplished by using the butt end (the handle side) of a pike pole to poke through the sheathing.[44]

Maintaining discipline is absolutely imperative when operating in buildings with knee walls, which are common in several types of multiple and private peaked-roof dwellings. Knee walls can be found wherever the half-story level has been finished to function as a living area. Steep pitches, dormered sections, and larger/full-size (egress) gable windows, as well as air conditioning units, window treatments (curtains and blinds), and fire escapes (for multiple dwellings, serving as the secondary means of egress) are all indicators that these voids are likely present. Knee walls are constructed to square off the sloping space, making it more functional and aesthetically pleasing. The large void spaces that are created, running parallel with the ridgepole, span the entire length of the roof on both sides. The top of the wall studs are attached to the underside of the roof rafters, leaving the rafter bays open and the voids behind the knee walls and above the ceiling connected. Also, due to the lack of fire stopping, particularly on older buildings, and/or utility chases or cutouts, fire in the stud channels below can extend vertically into these half-story void spaces (fig. 7–29).

Because of their size, the knee-wall voids can effectively conceal a tremendous volume of fire, heat, and smoke. If a firefighter unknowingly or impatiently opens up these walls or the ceiling before a charged handline is in position, the void's pressurized contents can rapidly extend out and fill the room, with the potential to almost instantly obscure visibility and deteriorate the conditions. Whether the connection is intentionally made or done naturally by the fire itself, the results can be catastrophic if necessary precautions are not taken. There have been numerous near misses and line-of-duty deaths attributed to extreme fire

Figure 7–29. The void spaces created by the knee walls and the ceiling
Courtesy: NBFD

behavior events involving knee walls. To limit fire extension and worsening conditions, opening up must be timed properly with extinguishment.

Closely coordinating vertical ventilation will also greatly aid in relieving the pressure built up within the voids and redirecting the fire and its by-products up and away from the crews operating below, enhancing visibility and tenability. There is some debate as to the best location of the hole placement. Making the cut at the highest point on the slope over the affected area, however, will vent both the knee-wall void and the void above the ceiling. Due to the steep pitches frequently encountered in these dwellings, it can be difficult to extend the cut out to roll at least two rafters, especially when working off of a roof ladder. Starting your cuts by the ridge will provide you with the freedom to extend the hole downward and increase the square footage, traveling toward the means of egress. Another tactic that can be employed to preserve the conditions is preemptively taking/opening the windows in the room before opening up the voids, which can also be done whenever pulling ceilings to expose concealed fire above. The open windows will immediately begin exhausting the smoke and the other by-products that will start banking down when the voids are penetrated, and even more so when the handline begins operating, helping to preserve visibility. Again, *a charged handline must be in position and ready to flow water prior to these openings being made* (fig. 7–30).

Greater scrutiny must be given when operating in half-stories and investigating for fire extension, especially when on the floor above the fire, and even more so in open stud-channel (balloon-frame) construction. Enlist all of your senses and do not rely solely on the temperature readings of your thermal imaging camera, which measures the heat of the *outside* surface. Watch for smoke issuing from cracks and seams (under pressure) and blistering or discolored paint or wallpaper. Listen for low rumbling sounds coming from behind walls or ceilings. With the back of your bare hand, lightly feel the surface in question to check for heat and see if you can comfortably hold it there. Do not forget about the

Figure 7–30. Concealed fire is exposed in the voids as the firefighters open up the ceilings and walls.
Courtesy: Lloyd Mitchell

all-important sixth sense—intuition (the "gut feeling"). If something does not feel right, it probably is not okay. Never leave any stone unturned. The potential benefits of being overly cautious or following a hunch, as long as it is justified and coordinated, far outweigh the consequences of an oversight.

Learning the Hard Way

We were the third-due engine company responding to a confirmed fire in a small, 1½-story bungalow (private dwelling with a second-floor living area). The first-arriving engine company found a small porch in the rear involved in fire and stretched a 1¾" handline. On our arrival, we were assigned the task of primary search on the first floor and to check for extension; the ladder company we arrived with was assigned the same task, but on the second floor. As we walked up to the house, no smoke could be seen, and it *appeared* the incident was essentially under control.

We assisted with forcing entry to the front door and conducted a search of the first floor with no smoke or fire encountered. With the search proving negative and no indication of extension, we proceeded to the second floor to link up with the ladder company. A mild smoke condition was present, but it was light and lazy, remaining just above head height. Scanning the rear bedroom with the thermal imaging camera yielded only a slight heat signature (125°F–135°F) on a small portion of the charlie-side wall (toward the bravo/charlie corner). On closer inspection, no signs of discoloration or blistering were present. At a cursory glance, it all *appeared* to be just residual heat and smoke that had migrated up through the stud channels from the exterior fire directly below.

The firefighter working in the irons position on the ladder company happened to catch a small wisp of smoke pushing from the corner of the window frame, just to the right of where the heat signature was detected (fig. 7–31). Sensing that something was not right, he called it to the attention of his officer, and a

Figure 7–31. Where the smoke was detected
Courtesy: NBFD

softball-sized inspection hole was opened in the wall. Flame and smoke instantly presented within the hole, prompting an immediate request for a handline to our location. *In less than a minute*, the small flame that appeared in the hole started blowtorching out, extending up the side of the wall. It began to spread across the ceiling, filling the room with smoke and banking down on top of us. In an effort to slow the fire's progress, the windows that had been opened earlier to vent the smoke were quickly shut. An attempt was made to control the bedroom door, only to discover *there was no door*.

With conditions rapidly deteriorating, the adjacent closet door was hastily removed as we retreated from the bedroom. The door was then held in place over the doorway to seal off the room, as the closet door was smaller than the opening. A heads-up member of my crew proceeded down the stairs to intercept the line being stretched, cutting down on the lag time and reducing the number of people in the already congested second-floor hallway. As we waited, we could see through the gaps of the door and watched as the fire continued to roll across the ceiling. The smoke layer dropped about halfway to the floor and began to move in the ominous wave-like motion. The line appeared just as the room was close to reaching flashover. When the line was advanced into the room, the fire was extinguished even more quickly than it had spread. The fire was contained to that room, and thankfully no one was injured. The outcome of this incident could have been dramatically different had it not been for the members reacting so swiftly and decisively.

The fire in the back porch below had communicated to the open stud channel running up to the room we were operating in on the second floor. The hidden pocket of fire quickly consumed the oxygen within the void space, becoming underventilated, and it erupted when fresh air was introduced by the inspection hole created during overhaul (fig. 7–32). This condition was further exacerbated by the open windows and absent bedroom door. The fire behavior encountered during this incident displays the volatility of underventilated fires and the diligence that must

Figure 7–32. The open stud channel
Courtesy: NBFD

be exercised when opening up *any* concealed space when checking for fire extension, especially when operating above the fire. It also highlights the importance of positional discipline and coordinating fireground operations.[45]

Guideline 10
Positioning and Egress

Ensure the integrity of the path being traveled and establish a secondary means of egress

When advancing throughout a building where structural members can be unknowingly compromised, it is imperative to verify the integrity of the path being taken. When traveling on the interior, stay low, but try to remain upright in a "tripod" position. By keeping one leg up, your weight is transferred onto the back leg. This allows the lead leg to probe/sound out while crawling, facilitating detection of any holes or weakened surfaces ahead. Being in this position not only allows firefighters to fall backward (opposite the affected area) in the event they lose their balance, but also keeps their heads up to monitor the fire conditions above.[46] The exception to this is when conditions require the firefighter to temporarily operate lower to the floor due to the heat, lack of visibility, or the configuration of the area being searched.

When operating on top of the building, a pike pole or rubbish hook must be used by the lead firefighter for sounding out the decking ahead before transitioning on and while walking across the roof. Firefighters carrying/operating a saw can roll the saw on its blade out in front of themselves as they reposition when using a rotary saw.[47] For operations on peaked roofs, especially on pitches greater than 6/12, a roof ladder should be deployed for added stability. Sheathing and environmental factors can make traction even more problematic, including metal decking, terra cotta and slate tiles, solar shingles, snow/ice, rain, and moss buildup. Once the cut operation begins, the sawdust produced on a wood-framed roof will further compound the issue. In some cases, the conditions may require operations to be conducted from an aerial ladder or off the platform of a bucket.

Whether operating inside or topside, it is critical to maintain a secondary means of egress. This can be as simple as identifying an alternative route such as a stairwell, fire escape, or an adjoining rooftop. If one is not present or is not ideal, another means can be created by the use of portable ladders and aerial devices. One of the best ways to avoid being cut off and requiring their use is to maintain your orientation and be mindful of the conditions, as well as your positioning, *at all times*. When operating without a handline, use the walls as your road map to navigate your way throughout the building.[48]

When placing ladders for access, avoid setting up in front of openings that have the potential to issue fire and directly impinge on them, cutting off that path of travel. If a wind condition is present (blowing across the face of the building being vented), position yourself so that the wind is at your back (the upwind side of the opening). Doing so will allow the fire and its by-products to exhaust away from you once the opening is created. If multiple openings are being made, start at the farthest point and work back towards the means of egress.[49] Whether you are operating on the interior or exterior of the building, be sure to keep the escape path at your back, allowing for an immediate retreat in the event conditions suddenly deteriorate (fig. 7–33).

Figure 7–33. The firefighter vents at the furthest point, working back toward the ladder.
Courtesy: Curt Hudson

MISSION-ORIENTED OPERATIONS

The best teams employ constant analysis of their tactics and measure their effectiveness so that they can adapt their methods and implement lessons learned....We would prioritize and execute to ensure the team concentrated our efforts on the most important facets, and we would focus our resources there.

—Commander Jocko Willink,
SEAL Team 3

As professionals, and most importantly, public servants, it is our *duty* to ensure we are organized and able to deploy with maximum effectiveness and efficiency. We must execute our operations in a manner that affords civilians and their property the highest level of protection and the best chance of survival. Doing so requires an in-depth assessment of our working environment, as well as our operational capacity and the full array of tactical options to address our specific needs. Only then can we develop the knowledge base and understanding to identify those that are truly compatible with our organization and the community we serve. This not only narrows down the selection process, thus streamlining operations, it inherently increases the chances of success on the fireground.

Ventilation is not a standalone tactic; it is a supportive function and a coordinated effort intended to augment firefighting operations and promote victim survivability. Because there are an infinite number of potential variables (some being unpredictable), we simply cannot dictate every aspect of the fireground, especially when it comes to ventilation. By venting tactically—managing the existing openings (when possible) and properly timing, placing, and sizing the creation of additional ones—we can maximize the level of support provided to both firefighters and victims alike.[1] Achieving the desired outcome requires that ventilation remains strictly task orientated (possessing a clear objective). It must be executed in cadence with the progress of fire attack and search and in alignment with the overall mission—preserving life and property (fig. 8–1).

We must never forget the oath that we solemnly took when we raised our right hand and began our journey in the fire service. Everything that we do, from administration to operations, must be with the best interests of the public's protection in the forefront of our minds. The Anchorage (AK) Fire Department proudly displays this motto: "I am not here for me, I am here for we, and we are here for *them*."[2] With our environment perpetually changing, technology advancing, and techniques refining, we need to be ardent students of our craft to uphold that promise. By constantly learning, drilling, and sharing, we can remain truly dedicated to not only the communities we swore to protect and the departments that we represent, but also to each other and to our families. Fire departments must foster a growth mindset and a mission-oriented culture, with a "bias for action." By continuously developing an "intellectual edge," along with a "brilliance in basic, fundamental skills," firefighters will possess the competencies and agility to act decisively and aggressively, regardless of the circumstances, while proactively adapting to the ever-changing conditions.[3]

Figure 8–1. The roof firefighter checks conditions in the rear as the outside vent firefighter operates on the fire escape below.
Courtesy: Matt Daly

Inadequate staffing, equipment, and training facilities have become a tragic reality for many fire departments across the country. Due to budget cuts and restrictions, especially those in urban communities, departments are forced to contend with these glaring deficiencies. Regardless of the situation, it is our obligation to safeguard the community to the best of our abilities. Achieving and maintaining this standard requires an organizational and individual commitment to preparedness and the relentless pursuit of excellence. Members need to believe in the significance of this undertaking and should be given the available resources and unwavering support to be successful. It is said that necessity is the mother of invention, so members *at all levels* should be empowered to apply their unique talents and personal experiences, thinking outside the box and pushing the limits of the normative constraints. We must embolden those individuals who are daring and passionate and who possess a genuine desire to advance their organization and all those within it. They are the very life blood of the fire service and the force multipliers that grant us the resiliency to triumph against all odds.

The fire service cannot afford to succumb to societal pressures to lower standards, embrace mediocrity, and tolerate complacency or indifference. The public demands the highest caliber of service and relies on our optimal performance to restore order in their greatest time of need. We must possess the discipline to put in the work necessary to achieve that standard *and the fortitude to uphold it*. On that inevitable day we are called on, there must be no hesitation or uncertainty, only steadfast action and *no regrets*.

GLOSSARY

above level. The floor above the one on which the seat of the fire is located. An above-level opening will almost exclusively function as an outlet, producing a unidirectional exhaust flow. Contrast with below level and on level.

adaptability. The capacity to adjust to a situation or a change in conditions, accomplished through proactive anticipation and/or reactive improvisation.

agility. The capacity to transition from one operation or task to another.

alligatoring. Deep char from extensive fire damage; resembles the skin texture of an alligator.

attic. The space between the decking of a peaked roof and the top-floor ceiling.

autoexposure. The extension of fire via the exterior of a building from a fire originating on a lower level.

backdraft. When a fire occurs within a closed compartment and consumes most of the available oxygen, the heat within the space may continue to produce flammable gases, which are heated above their ignition temperature. When a supply of oxygen is then introduced and mixes with the fire gases—entering into their flammable range—an explosive event (*deflagration*) can occur, causing the gases to ignite with tremendous force. See also deflagration.

below level. The floor beneath the one on which the seat of the fire is located. A below-level opening will almost exclusively function as an air flow inlet, producing a unidirectional flow. Contrast with above level and on level.

Bernoulli's principle. An increase in the speed of an incompressible fluid flowing within a vessel, occurring simultaneously with a decrease in the static pressure of the fluid, that results from a region being reduced in diameter.

bidirectional flow. Gases flowing in opposing directions. The hot combustion gases in the upper portion are migrating away from the fire and the cool fresh air is traveling toward it in the lower portion. When occurring at a ventilation opening, it results in a less efficient air exchange since the intaking and exhausting gases are competing for the same area.

British thermal unit (Btu). The US customary unit of measurement for the amount of heat needed to raise 1 pound of water 1° Fahrenheit.

bulkhead. A structure built over a stairway, elevator, dumbwaiter shaft, or other building facility on the roof.

calorimetry. The process of using a calorimeter to measure the amount of heat released during a chemical reaction.

can. A 2.5-gallon pressurized water extinguisher (typically carried by the junior firefighter of a ladder company's inside crew).

carbonization. The production of carbon particulate as a fuel undergoes pyrolysis.

cockloft. The area between the top-floor ceiling and the underside of the roof deck; typically associated with a flat roof.

cockloft explosion. An explosion within a cockloft or attic space that typically occurs after an influx of air is introduced. When the top-floor ceiling is breached or access is otherwise gained to that area, an explosion may result from the previously confined, fuel-rich environment (known as a *backdraft*). This can occur if the fire is in the cockloft and begins to decay or even if the fire is in an adjacent/interconnected space and the by-products migrate and accumulate in the cockloft. A blast of pressure caused by the sudden ignition of the fire gases (*deflagration*) once they reach the proper oxygen/fuel mixture and a source of ignition (known as a *smoke explosion*) can violently collapse the ceiling. See also backdraft, smoke explosion, and deflagration.

commander's intent. The rationale of why an assignment is being given and the end state that is expected.

convection. The transfer of heat by a liquid or gas.

crooked-lean position. The nozzle firefighter leaning back onto the back-up firefighter for support in order to flow and move a handline on the approach to the seat of the fire.

decay stage. The stage of fire development characterized by a depletion of the fuel load consumed during the combustion process. It can also occur prematurely if the oxygen supply is insufficient. The decay stage results in reduced temperature and pressure in either case. Contrast with incipient stage and growth stage.

decentralized command. When subordinates possess a clear definition of what the mission and objectives are, along with the operational parameters and boundaries, and are equipped with the requisite knowledge and skill sets to be successful, they can be given the freedom to make tactical-level decisions based on the tacit knowledge of the situation they are encountering and any changing conditions.

deflagration. Rapid combustion propagating through hot gases at subsonic speed; capable of producing an explosive effect when occurring within an enclosed space. See also backdraft and smoke explosion.

diagnostic cut. A kerf or triangular inspection cut made in the roof as a means of checking for fire in the attic/cockloft space and to determine the level of *fire extension* (identifying the boundaries of the fire area) and the construction. See also inspection hole and kerf cut.

draft stop. A secondary skylight located beneath the main skylight on the roof, flush with the ceiling below. It is designed to prevent the conditioned air within the living space from entering the skylight well.

exothermic reaction. A chemical process that results in the production of energy in the form of heat and/or light.

fire. The rapid oxidation of a material through the process of combustion.

fire plume. A column of combustion by-products rising upward above a fire; caused by the buoyancy of the heated gases. As the heated gases rise, cooler air is drawn into the base of the fire, producing a negative pressure and entraining air from adjoining compartments.

fire tetrahedron. A visual representation of the mutual relationship between the three fundamental parts of the combustion process—fuel, oxygen, and heat—that collectively form a chemical chain reaction.

fire triangle. A visual representation of the mutual relationship between the three fundamental parts of the combustion process: fuel, oxygen, and heat.

flashover. A transition in the development of a compartment fire when surfaces exposed to thermal radiation from fire gases reach ignition temperature in rapid succession, resulting in rapid fire spread and total involvement of the compartment.

flow path. The volume of space within a compartment where gases travel. Flames, heat, and smoke migrate from a higher pressure source (i.e., the fire) toward areas of lower pressure, both within the structure and to the outside atmosphere, while cool fresh air is inversely drawn in to replace them. The flow path occurs via entryways, window openings, stairways, and roof structures.

fuel-limited fire. A fire in which the heat release rate and fire growth are controlled by the fuel load given that an adequate supply of oxygen is readily available for combustion.

fuel-rich fire. See ventilation-limited fire.

fully developed stage. The stage of fire development that has reached its peak heat release rate within a compartment. This usually occurs after a flashover, resulting in floor-to-ceiling burning within the compartment. See also flashover.

Gay-Lussac's law. The description of the relationship between pressure and temperature in a constant volume and a fixed mass.

growth stage. The stage of fire development when the heat release rate from an incipient fire has increased to the point where adjacent fuel sources are pyrolyzed by radiant heat transferred from the fire and the combustion products. A second growth stage can occur when a ventilation-limited fire—in the (premature) decay phase—receives a fresh supply of oxygen. Contrast with incipient stage and decay stage.

heat release rate (HRR). The rate at which energy is generated in the combustion process, which is determined by the characteristics of the fuel and regulated by the supply of oxygen.

incipient stage. The early stage of fire development where the fire's progression is limited to a single fuel source and the thermal hazard is localized to the area of the burning material. Contrast with growth stage and decay stage.

initiative. The ability to assess and initiate operations independently and decisively in response to changes in the environment and the situation and the subsequent need to achieve an objective and/or support the overall mission.

inspection hole. A small hole placed into a wall or ceiling to check for concealed fire in the void spaces or a triangular cut made in the roof to check for extension into the attic/cockloft space, as well as to determine its extent (i.e., the boundaries of the fire). See also diagnostic cut.

intuition. The ability to inherently comprehend a situation or predict a future event and to act decisively. Rooted in the brain's ability to subconsciously pattern match; based on past experiences, supplemental education/training, and the level of situational awareness exhibited.

irons firefighter. The firefighter (typically assigned to the inside team of a ladder company) who is assigned to carry a flat-head axe and a Halligan bar (referred to as "the irons" when married together).

joule (J). The SI unit of measurement for the energy transferred to an object when a force of 1 newton acts on that object in the direction of the force's motion through a distance of 1 meter.

kerf cut. A type of diagnostic cut created by dropping the blade of the saw into the roof decking to create a narrow inspection opening (the width or "kerf" of the blade) to identify the presence of fire in the space below (i.e., the attic/cockloft).

knee walls. A short wall, usually less than 3' high, used to support rafters in a peaked-roof and/or to frame out the attic, creating a finished/livable space.

laminar smoke. Smoke discharging in a tight, streamlined fashion; evidence that the building and its contents are still capable of absorbing the heat being generated by the fire.

louvers. A wood or vinyl wall vent consisting of overlapping horizontal slats, typically found on the gable ends of peaked-roof dwellings to allow for natural air flow within the attic space.

mindset. The attitude and beliefs held by a person (influenced by past experiences and education/training) that determine how that person will interpret and react to his or her environment and the situation.

mushrooming. As the thermal updraft causes the smoke to rise, it will collect along the ceiling until it reaches a horizontal barrier, causing it to then begin to bank back down.

neutral plane. The point of thermal balance when observing smoke exhausting out of an opening from the exterior.

on level. The floor on which the seat of the fire is located. An on-level opening will typically function as both an inlet and an outlet, producing a bidirectional flow. Contrast with above level and below level.

pipe chase. An unobstructed vertical channel, starting in the basement and extending up to the attic/cockloft space, which houses the plumbing/conduit for the utilities; a notable avenue for fire spread.

potential energy. The defined heat capacity of a particular fuel, which is determined by its chemical makeup.

pressure gradient. The physical quantity that describes in which direction and at what rate the pressure increases.

pyrolysis. The chemical decomposition of a fuel due to heat.

RADE loop. A progressive/cyclical decision-making model that identifies the variables and contributing factors, as well as the evolving size-up process: (1) Read the conditions; (2) Anticipate the progression; (3) Determine the needs, and (4) Execute the tactics.

radiation. The transfer of heat through electromagnetic waves.

radiation feedback. The heat energy of the fire being radiated back between the fuels and the surfaces within the compartment.

rain roof. A sloping or peaked roof built over an old, flat roof deck as a means of waterproofing or repair.

reflex time. The required or elapsed time between the issuance of an order and when the desired effect or end state is achieved.

returns. The walls between the roof and the top-floor ceiling that form the enclosure of a scuttle or skylight opening.

rollover. A phenomenon that occurs in the growth stage when sufficient fuel, heat, and oxygen are available to allow flame spread in the upper hot-gas layer; an imminent precursor to a flashover. See also flashover and growth stage.

sash. The metal, wood, or vinyl window supports that frame out the glass (often referring to the center crossmember).

scuttle. An access opening in the roof with a hatch or removable cover. Typically located at the top of stairwells or in utility closets off of the common hallway on the top floor; often possessing a fixed ladder leading up to the opening.

sensory input. The stimuli being received through the five senses—sight, smell, hearing, touch, and taste—and then processed/interpreted by the brain.

shaftways. Vertical spaces in multiple-story/multiple-occupancy dwellings, intended to bring sunlight and air into the rooms of the apartments through windows. Located in the interior of the building, shaftways typically extend from the ground level to the roofline. In some cases, the shafts start at the first- or second-floor level and are then known as *blind shafts*.

situational awareness. The continuous assessment of your environment and the situation to determine the best course of action.

smoke. The by-product of incomplete combustion; unburned fuel. A combination of airborne solid particulates, aerosols, and gases/vapors emitted when a fuel source undergoes pyrolysis. Smoke contains a mixture of toxic gases that can readily achieve fatal concentrations during the course of a fire.

smoke explosion. A sudden ignition of fire gases in a confined space, typically remote from the main body of fire. The by-products of combustion migrate through void spaces within the building and accumulate within adjacent/interconnected spaces. Once the fire gases are within their flammable range (i.e., proper oxygen/fuel mixture) and an ignition source is introduced (which may be the fire itself), the explosive ignition (*deflagration*) of those gases results in a potentially significant blast of pressure. See also deflagration.

smoke tunneling. After a ventilation opening has been created that allows the smoke to exhaust and the thermal balance to lift, a void is created below, allowing fresh air to intake toward the fire.

tempo. The speed of action (tactics) relative to a problem set (the fire and the environment).

tenability. The conditions within a specific area that determine the extent to which operations can be carried out and the survivability of potential victims.

thermal balance. The horizontal dividing line that separates the heat and smoke rising/collecting above and the cooler air remaining below.

thermal updraft. The upward movement of air as it becomes heated due to its increased buoyancy.

Thornton's rule. For each unit of oxygen consumed for combustion, there is a fairly constant heat output among common organic fuels: 13.1 MJ of heat/energy per 1 kg of oxygen.

throat. The narrow area of a wing-type building that connects the larger sections (e.g., U- and H-shaped); making it the most advantageous position to place a trench cut.

trench cut. A 3' wide defensive, vertical ventilation opening cut from the edge of the roof to the opposite end to create a firebreak and halt lateral fire spread in the cockloft of a flat-roof structure. The trench cut is utilized when the primary ventilation hole over the main body of fire has been overwhelmed and expanding it is no longer a viable option.

turbulent smoke. Smoke exhausting erratically (chugging and twirling); evidence that the compartment has become heat saturated to capacity. The excessive buildup of heat and radiation feedback causes the gases within the smoke to rapidly expand, but the gases lack an exhaust outlet of sufficient size; a sign of impending flashover. See also flashover.

unidirectional flow. Gases flowing in a single direction. When occurring at a ventilation opening, it results in the most efficient air exchange, functioning as either an exclusive intake point (drawing in fresh air) or exhaust point (discharging hot combustion gases).

upper register. The ceiling and the topmost portion of the walls; the area where the heat and smoke accumulate during a fire.

vent-point ignition. The sudden ignition of fuel-rich, heated smoke as it exhausts from an outlet and mixes with the outside atmosphere, leaning out and achieving the proper air-to-fuel mixture.

ventilation. A vital fireground task intended to systematically remove the by-products of combustion and extinguishment from a structure and replace them with fresh air (through the creation and management of new and existing openings) to support firefighting efforts and victim survivability.

ventilation for access. Creating an opening via a doorway or a window for the purposes of gaining entry to the building in order to initiate extinguishment and/or search operations, or penetrating the walls of ceilings to gain access to the void spaces within the buildings.

ventilation for extinguishment. Creating an exhaust opening for the purposes of relieving conditions for the attacking engine company and aiding in the confinement of the fire as extinguishment is taking place; also known as *venting for fire*.

ventilation for search. Creating an opening for the purposes of relieving conditions for the search crews and any trapped victims; also known as *venting for life*.

ventilation-induced flashover. A flashover initiated by the sudden introduction of oxygen to a ventilation-limited fire. The hot, fuel-rich environment is leaned out by the influx of fresh air, bringing the fire gases within their flammable range. See also flashover.

ventilation-limited fire. A fire in which the heat release rate and fire growth are regulated by the available oxygen within the space; also known as a *fuel-rich fire*.

ventilation profile. Identifying the fire location, the wind velocity, and the openings present within a building that function as inlets and/or outlets. These determine the type of flow and the pathway of the ventilation, indicating the migration of the fire and its by-products, as well as the flow of fresh air.

watt (W). The SI unit of measurement for power; equivalent to 1 joule per second.

NOTES

The problem with being too busy to read is that you learn by experience (or by your men's experience), i.e., the hard way. By reading, you learn through others' experiences, generally a better way to do business, especially in our line of work where the consequences of incompetence are so final for young men.

—General James Mattis, USMC

Preface

1. A. M. Gray, *FMFM-1: Warfighting* (Washington, DC: US Marine Corps, US Department of the Navy, 1989).
2. Gray, *FMFM-1: Warfighting*.

Chapter 1

1. Michael Asken, *Warrior Mindset: Mental Toughness Skills for a Nation's Defenders and Performance Psychology Applied to Combat* (Warrior Science Publications, 2009).
2. Aaron Fields, personal interview, Fire Engineering Blog Talk Radio (March 19, 2017).
3. Dave Grossman, *On Combat: The Psychology and Physiology of Deadly Conflict in War and in Peace* (Warrior Science Publications, 2008).
4. Grossman, *On Combat*; Robert Brown, *Smart 4 Life* (lecture, 36th Annual James Curran Memorial Seminar, Bureau of Training, Fire Department New York, New York City, April 16, 2016); Robert Brown, "Auditory Exclusion and How the Sympathetic Response Affects Our Operational Performance on the Fireground," *With New York Firefighters* 2 (2016); Robert Brown, "Smart Venting Using the SA Cycle," *With New York Firefighters* 2 (2014); Jonathan Fader, *Life as Sport: What Top Athletes Can Teach You About How to Win* (Boston: Da Capo Press, 2016), Eric Nurnberg and Michael Asken, *Fire Psyche: Mental Toughness and the Valor Mindset for the Fireground* (Mind Sighting, 2014).
5. Stephen Kerber, *Top 20 Tactical Considerations from Firefighter Research* (Columbia, MD: UL–Firefighter Safety Research Institute, 2016).

Chapter 2

1. IFSTA, *Structural Fire Fighting: Initial Response Strategy and Tactics*, 2nd ed. (Stillwater, OK: IFSTA, 2017)
2. David Fornell, *Fire Stream Management Handbook* (Saddle Brook, NJ: Fire Engineering Books & Videos, 1991).
3. Ed Hartin, "Thornton's Rule," CFBT-US.com (2009).
4. Vytenis Babrauskas and R. D. Peacock, "Heat Release Rate: The Single Most Important Variable in Fire Hazard," *Fire Safety Journal* 18, no. 3 (1992): 255–272, https://doi.org/10.1016/0379-7112(92)90019-9.
5. James Braidwood, *On the Construction of Fire-Engines and Apparatus, the Training of Firemen, and the Method of Proceedings in Cases of Fire* (London: Bell and Bradfute, 1830).
6. W. M. Thornton, "The Relation of Oxygen to the Heat of Combustion of Organic Compounds," *London, Edinburgh, and Dublin Philosophical Magazine and Journal of Science* 33, 194 (1917); and W. F. Crapo, "Thornton's Rule: Anticipating the Heat-Release Rate," *Fire Engineering* 164, no. 5 (2011).
7. Clayton Huggett, "Estimation of Rate of Heat Release by Means of Oxygen Consumption Measurements," *Fire and Materials* 4, no. 2 (June 1980): 61–65, https://doi.org/10.1002/fam.810040202.
8. Huggett, "Estimation of Rate of Heat Release by Means of Oxygen Consumption Measurements"; and Hyeong-Jin Kim and David Lilley, "Heat Release Rates of Burning Items in Fires," AIAA 2000-0722 (Stillwater, OK: American Institute of Aeronautics and Astronautics (AIAA), 2000).
9. Paul Grimwood, *Euro Firefighter 2: 6,701 Building Fires* (West Yorkshire, UK: D&M Heritage Press, 2017).
10. Hartin, *Thornton's Rule*.
11. Hartin, *Thornton's Rule*; Grimwood, *Euro Firefighter 2*; Stefan Svensson, *Fire Ventilation* (Swedish Rescue Services Agency, 2005); Dan Madrzykowski, "Fire Dynamics for Structural Firefighting," Presentation, UL–Firefighter Safety Research Institute (2017); Jason N. Vestal and Eric A. Bridge, "A Quantitative Approach to Selecting Nozzle Flow Rate and Stream, Part I," *Fire Engineering*, 163, no. 10 (October 1, 2010), https://www.fireengineering.com/2010/10/01/208182/a-quantitative-approach-to-selecting-nozzle-flow-rate-and-stream-part-1/; and Jason N. Vestal and Eric A. Bridge, "A Quantitative Approach to Selecting Nozzle Flow Rate and Stream, Part II," *Fire Engineering*, 164, no. 1 (January 1, 2011), https://www.fireengineering.com/2011/01/01/250276/a-quantitative-approach-to-selecting-nozzle-flow-rate-and-stream-part-2/.
12. Ed Hartin, "Influence of Ventilation in Residential Structures: Tactical Implications," Part 1, CFBT-US.com (2011).
13. Keith Stakes, et al., *Analysis of the Coordination of Suppression and Ventilation in Multi-Family Dwellings* (Columbia, MD: UL–Firefighter Safety Research Institute, 2020), https://ulfirefightersafety.org/docs/Coord_Tactics_Multi_Family.pdf.

14. Aaron Fields, (lecture, The Nozzle Forward Training Program, May 11, 2016).
15. Vincent Dunn, *Strategy of Firefighting* (Tulsa, OK: Fire Engineering Books & Videos, 2007).
16. Stakes, Bryant, Dow, Regan, and Weinschenk, *Analysis of the Coordination of Suppression and Ventilation in Multi Family Dwellings*.
17. Nicholas Papa, "The Impact of Ventilation at Structural Fires: Breaking Down the Fire Triangle," *Fire Engineering* 171, no. 4 (April 1, 2018), https://www.fireengineering.com/2018/04/01/252227/the-impact-of-ventilation-at-structural-fires-breaking-down-the-fire-triangle/#gref; and John Norman, "Store Fire Tactics," (lecture, Lt. Joseph DiBernardo Memorial Seminar, Stony Brook University, Stony Brook, New York, November 5, 2016).
18. Peter Van Dorpe, *The New Fire Fight* (Tulsa, OK: Fire Engineering Books & Videos, 2018), DVD.
19. David Dodson, "The Art of Reading Smoke," *Fire Engineering* (June 27, 2014), https://www.fireengineering.com/2014/06/27/236553/david-dodson-the-art-of-reading-smoke/; and David Dodson, *The Art of Reading Smoke*, Vols. 1–3 (Tulsa, OK: Fire Engineering Books & Videos, 2007).
20. Dodson, "The Art of Reading Smoke"; and Dodson, *The Art of Reading Smoke*.
21. Dodson, "The Art of Reading Smoke"; and Dodson, *The Art of Reading Smoke*.
22. Dodson, "The Art of Reading Smoke"; and Dodson, *The Art of Reading Smoke*.
23. Dodson, "The Art of Reading Smoke"; and Dodson, *The Art of Reading Smoke*.
24. Fire Department of New York (FDNY), "Ventilation," *Firefighting Procedures* 1, no. 10 (2013); and Fire Department of New York (FDNY), "Ventilation," *Firefighting Procedures* 1, no. 10 (2019).
25. Dodson, "The Art of Reading Smoke"; and Dodson, *The Art of Reading Smoke*.
26. Norman, "Store Fire Tactics"; and Dodson, "The Art of Reading Smoke."
27. Dodson, "The Art of Reading Smoke"; and Dodson, *The Art of Reading Smoke*.
28. Dodson, "The Art of Reading Smoke"; and Dodson, *The Art of Reading Smoke*.
29. Dodson, "The Art of Reading Smoke"; and Dodson, *The Art of Reading Smoke*.

Chapter 3

1. Ed Hartin, "Influence of Ventilation in Residential Structures: Tactical Implications," Part 1 (CFBT-US, 2011).
2. Jim McCormack, "The Modern Fireground," *Fire Department Training Network*, 22, no. 11 (2018): 10–12; and Stephen Kerber, *The Impact of Ventilation on Fire Behavior in Legacy and Contemporary Residential Construction* (Columbia, MD: UL–Firefighter Safety Research Institute, 2010).
3. Kerber, *The Impact of Ventilation on Fire Behavior in Legacy and Contemporary Residential Construction*; and Stephen Kerber, *Study of the Effectiveness of Fire Service Vertical Ventilation and Suppression Tactics in Single Family Homes*

(Columbia, MD: UL–Firefighter Safety Research Institute, 2013), https://ulfirefightersafety.org/docs/UL-FSRI-2010-DHS-Report_Comp.pdf.
4. Jack Regan, Julie Bryant, and Craig Weinschenk, *Analysis of the Coordination of Suppression and Ventilation in Single-Family Homes* (Columbia, MD: UL–Firefighter Safety Research Institute, 2020).
5. Kerber, *The Impact of Ventilation on Fire Behavior in Legacy and Contemporary Residential Construction*; and Kerber, *Study of the Effectiveness of Fire Service Vertical Ventilation and Suppression Tactics in Single Family Homes*.
6. Keith Stakes, Julie Bryant, Nicholas Dow, Jack Regan, and Craig Weinschenk, *Analysis of the Coordination of Suppression and Ventilation in Multi-Family Dwellings* (Columbia, MD: UL–Firefighter Safety Research Institute, 2020), https://ulfirefightersafety.org/docs/Coord_Tactics_Multi_Family.pdf.
7. Regan, Bryant, and Weinschenk, *Analysis of the Coordination of Suppression and Ventilation in Single-Family Homes*.
8. James Braidwood, *Fire Prevention and Extinction* (London: Bell and Bradfute, 1866).
9. Kerber, *The Impact of Ventilation on Fire Behavior in Legacy and Contemporary Residential Construction*; and Kerber, *Study of the Effectiveness of Fire Service Vertical Ventilation and Suppression Tactics in Single Family Homes*.
10. Emanuel Fried, *Fireground Strategies* (Saddlebrook, NJ: Fire Engineering Books & Videos, 1972).
11. Aaron Fields, Personal interview, Fire Engineering Blog Talk Radio (March 19, 2017).
12. Kerber, *The Impact of Ventilation on Fire Behavior in Legacy and Contemporary Residential Construction*; and Kerber, *Study of the Effectiveness of Fire Service Vertical Ventilation and Suppression Tactics in Single Family Homes*.
13. Kerber, *The Impact of Ventilation on Fire Behavior in Legacy and Contemporary Residential Construction*; and Kerber, *Study of the Effectiveness of Fire Service Vertical Ventilation and Suppression Tactics in Single Family Homes*.
14. Lloyd Layman, *Attacking and Extinguishing Interior Fires* (Boston: National Fire Protection Association, 1955).
15. Kerber, *The Impact of Ventilation on Fire Behavior in Legacy and Contemporary Residential Construction*; and Kerber, *Study of the Effectiveness of Fire Service Vertical Ventilation and Suppression Tactics in Single Family Homes*.
16. Braidwood, *Fire Prevention and Extinction*.
17. Regan, Bryant, and Weinschenk, *Analysis of the Coordination of Suppression and Ventilation in Single-Family Homes*.
18. Aaron Fields, personal interview; Stakes, Bryant, Dow, Regan, and Weinschenk, *Analysis of the Coordination of Suppression and Ventilation in Multi-Family Dwellings*; Regan, Bryant, and Weinschenk, *Analysis of the Coordination of Suppression and Ventilation in Single-Family Homes*; and Robin Zevotek and Stephen Kerber, *Study of the Effectiveness of Fire Service Positive Pressure Ventilation during Fire Attack in Single Family Homes Incorporating Modern Construction Practices* (Columbia, MD: UL–Firefighter Safety Research Institute, 2016), https://ulfirefightersafety.org/docs/Positive_Pressure_Ventilation_Report_Website.pdf.
19. Craig Weinschenk and Robin Zevotek, *Exploratory Analysis of the Impact of Ventilation on Strip Mall Fires* (Columbia, MD: UL–Firefighter Safety Research Institute, 2020).

20. Weinschenk and Zevotek, *Exploratory Analysis of the Impact of Ventilation on Strip Mall Fires.*
21. Edward McAniff, *Strategic Concepts of Fire Fighting* (Saddlebrook, NJ: Fire Engineering Books & Videos, 1974).
22. Robin Zevotek, "Research Corner: Gaining Access to Ventilation-Limited Fires," *Firehouse* (April 1, 2017), https://www.firehouse.com/home/article/12305355/research-corner-gaining-access-to-ventilationlimited-fires.
23. Hossein Davoodi, *Confinement of Fire in Buildings* (Quincy, MA: National Fire Protection Agency, 2008).
24. Weinschenk and Zevotek, *Exploratory Analysis of the Impact of Ventilation on Strip Mall Fires.*
25. Stephen Kerber, *Top 20 Tactical Considerations from Firefighter Research* (Columbia, MD: UL–Firefighter Safety Research Institute, 2016); and Stefan Svensson, *Fire Ventilation* (Swedish Rescue Services Agency, 2005).
26. Kerber, *The Impact of Ventilation on Fire Behavior in Legacy and Contemporary Residential Construction*; Kerber, *Study of the Effectiveness of Fire Service Vertical Ventilation and Suppression Tactics in Single Family Homes*; and Regan, Bryant, and Weinschenk, *Analysis of the Coordination of Suppression and Ventilation in Single-Family Homes.*
27. Regan, Bryant, and Weinschenk, *Analysis of the Coordination of Suppression and Ventilation in Single-Family Homes.*
28. Regan, Bryant, and Weinschenk, *Analysis of the Coordination of Suppression and*
29. *Ventilation in Single-Family Homes,* 395.
30. Regan, Bryant, and Weinschenk, *Analysis of the Coordination of Suppression and Ventilation in Single-Family Homes.*
31. Regan, Bryant, and Weinschenk, *Analysis of the Coordination of Suppression and Ventilation in Single-Family Homes.*
32. Regan, Bryant, and Weinschenk, *Analysis of the Coordination of Suppression and Ventilation in Single-Family Homes.*
33. Kerber, *Top 20 Tactical Considerations from Firefighter Research*; and Zevotek, "Research Corner: Gaining Access to Ventilation-Limited Fires."
34. Regan, Bryant, and Weinschenk, *Analysis of the Coordination of Suppression and Ventilation in Single-Family Homes.*
35. Regan, Bryant, and Weinschenk, *Analysis of the Coordination of Suppression and Ventilation in Single-Family Homes.*
36. Regan, Bryant, and Weinschenk, *Analysis of the Coordination of Suppression and Ventilation in Single-Family Homes.*

Chapter 4

1. Fire Department of New York (FDNY), "Ventilation," *Firefighting Procedures* 1, no. 10 (2013); and Fire Department of New York (FDNY), "Ventilation," *Firefighting Procedures* 1, no. 10 (2019).

2. Stephen Kerber, *Top 20 Tactical Considerations from Firefighter Research* (Columbia, MD: UL–Firefighter Safety Research Institute, 2016).
3. Dennis LeGear, Facebook post (May 28, 2017), https://www.facebook.com/photo.php?fbid=435139000194824&set=pcb.435139030194821&type=3&theater.
4. Jack Regan, Julie Bryant, and Craig Weinschenk, *Analysis of the Coordination of Suppression and Ventilation in Single-Family Homes* (Columbia, MD: UL–Firefighter Safety Research Institute, 2020).
5. Lloyd Layman, *Fire Fighting Tactics* (Boston, MA: National Fire Protection Association, 1953).
6. Kerber, *Top 20 Tactical Considerations from Firefighter Research*.
7. Fire Department New York (FDNY), *Ladders 4: Operations at Private Dwellings* (NY: FDNY, 2013).
8. John Ceriello, comment on Facebook post by Kevin Burns, "How a Meme about Ventilation Went 'FDNY Viral,'" (2017). The information in the post has been summarized and is available from the Fire Engineering Training Community, Fire Engineering.com (September 25, 2017), https://community.fireengineering.com/m/blogpost?id=1219672%3ABlogPost%3A641230.
9. John Norman, *The Fire Officer's Handbook of Tactics*, 4th ed. (Tulsa, OK: Fire Engineering Books & Videos, 2012).
10. David Fornell, *Fire Stream Management Handbook* (Saddlebrook, NJ: Fire Engineering Books & Videos, 1991).
11. Jocko Willink and Leif Babin, *Extreme Ownership: How Navy Seals Lead and Win* (New York: St. Martin's Press, 2016).
12. Norman, *The Fire Officer's Handbook of Tactics*.
13. D. H. Berger, *MCDP-7: Learning* (Washington, DC: US Marine Corps, Department of the Navy, 2020), https://www.marines.mil/Portals/1/Publications/MCDP%207.pdf?ver=2020-03-03-111011-120.
14. Craig Weinschenk and Robin Zevotek, *Exploratory Analysis of the Impact of Ventilation on Strip Mall Fires* (Columbia, MD: UL–Firefighter Safety Research Institute, 2020).
15. Stephen Kerber, *The Impact of Ventilation on Fire Behavior in Legacy and Contemporary Residential Construction* (Columbia, MD: UL–Firefighter Safety Research Institute, 2010); and Stephen Kerber, *Study of the Effectiveness of Fire Service Vertical Ventilation and Suppression Tactics in Single Family Homes* (Columbia, MD: UL–Firefighter Safety Research Institute, 2013), https://ulfirefightersafety.org/docs/UL-FSRI-2010-DHS-Report_Comp.pdf; and Regan, Bryant, and Weinschenk, *Analysis of the Coordination of Suppression and Ventilation in Single-Family Homes*.
16. Vincent Dunn, *Strategy of Firefighting* (Tulsa, OK: Fire Engineering Books & Videos, 2007).
17. FDNY, *Ladders 4: Operations at Private Dwellings*.
18. Norman, *The Fire Officer's Handbook of Tactics*.
19. Vincent Dunn, *Safety and Survival on the Fireground* (Saddlebrook, NJ: Fire Engineering Books & Videos, 1992).

20. Regan, Bryant, and Weinschenk, *Analysis of the Coordination of Suppression and Ventilation in Single-Family Homes.*
21. Regan, Bryant, and Weinschenk, *Analysis of the Coordination of Suppression and Ventilation in Single-Family Homes.*
22. Nicholas Papa, "Venting for Extinguishment in Peak-Roof Dwellings," *Fire Engineering* (October 2019): 67–72, https://digital.fireengineering.com/fireengineering/october_2019/MobilePagedReplica.action?pm=2&folio=66#pg68.
23. Craig Weinschenk, Keith Stakes, and Robin Zevotek, *Impact of Fire Attack Utilizing Interior and Exterior Streams on Firefighter Safety and Occupant Survival: Air Entrainment* (Columbia, MD: UL–Firefighter Safety Research Institute, December 2017), https://ulfirefightersafety.org/docs/DHS2013_Part_II_Air_Entrainment.pdf.
24. Keith Stakes, Julie Bryant, Nicholas Dow, Jack Regan, and Craig Weinschenk, *Analysis of the Coordination of Suppression and Ventilation in Multi-Family Dwellings* (Columbia, MD: UL–Firefighter Safety Research Institute, 2020), https://ulfirefightersafety.org/docs/Coord_Tactics_Multi_Family.pdf.
25. Stakes, Bryant, Dow, Regan, and Weinschenk, *Analysis of the Coordination of Suppression and Ventilation in Multi-Family Dwellings.*
26. Robin Zevotek and Stephen Kerber, *Study of the Effectiveness of Fire Service Positive Pressure Ventilation during Fire Attack in Single Family Homes Incorporating Modern Construction Practices* (Columbia, MD: UL–Firefighter Safety Research Institute, 2016), https://ulfirefightersafety.org/docs/Positive_Pressure_Ventilation_Report_Website.pdf.
27. Zevotek and Kerber, *Study of the Effectiveness of Fire Service Positive Pressure Ventilation during Fire Attack in Single Family Homes Incorporating Modern Construction Practices.*
28. Zevotek and Kerber, *Study of the Effectiveness of Fire Service Positive Pressure Ventilation during Fire Attack in Single Family Homes Incorporating Modern Construction Practices.*

Chapter 5

1. A. M. Gray, *FMFM-1: Warfighting* (Washington, DC: US Marine Corps, US Department of the Navy, 1989); Paul Grimwood, *Euro Firefighter 2: 6,701 Building Fires* (West Yorkshire, UK: D&M Heritage Press, 2017); C. C. Krulak, *MCDP 1-3: Tactics* (Washington, DC: US Marine Corps, US Department of the Navy, 1997), https://www.defenceweb.co.za/wp-content/uploads/Repository/US-Marine-Corps-Manuals/MCDP1-3-Tactics.pdf; Robert Coram, *Boyd: The Fighter Pilot Who Changed the Art of War* (New York: Back Bay Books, 2002); Gary Klein, *The Power of Intuition: How to Use Your Gut Feelings to Make Better Decisions at Work* (New York: Crown Publishing, 2003); and Gary Klein, *Sources of Power: How People Make Decisions* (Cambridge, MA: MIT Press, 1998).
2. Grimwood, *Euro Firefighter 2.*

3. Aaron Fields, "The Attacking Handline," (lecture, The Nozzle Forward Training Program, 2019).
4. Vincent Dunn, *Safety and Survival on the Fireground* (Saddle River, NJ: Fire Engineering Books & Videos, 1992).
5. Gray, *FMFM-1: Warfighting*; and Peter Blaber, *The Mission, the Men, and Me: Lessons from a Former Delta Force Commander* (New York: Penguin Group, 2008).
6. Coram, *Boyd: The Fighter Pilot Who Changed the Art of War*.
7. Fire Department of New York (FDNY), "Ventilation," *Firefighting Procedures* 1, no. 10 (2013); and Fire Department of New York (FDNY), "Ventilation," *Firefighting Procedures* 1, no. 10 (2019).
8. Jason Brezler and Thomas Richardson, "Fireground Tempo," (lecture, 37th Annual James Curran Memorial Seminar, Bureau of Training, Fire Department New York, New York City, April 22, 2017).
9. Brezler and Richardson, "Fireground Tempo."
10. Klein, *The Power of Intuition*.
11. Klein, *The Power of Intuition*.
12. Gray, *FMFM-1: Warfighting*.
13. Klein, *The Power of Intuition*; and Malcolm Gladwell, *Blink: The Power of Thinking without Thinking* (Boston, MA: Back Bay Books, 2007).
14. Klein, *The Power of Intuition*.
15. William McRaven, *Spec Ops: Case Studies in Special Operations Warfare: Theory and Practice* (New York: Presidio Press, 2009)
16. Krulak, *MCDP 1-3: Tactics*.
17. Krulak, *MCDP 1-3: Tactics*.
18. Coram, *Boyd: The Fighter Pilot Who Changed the Art of War*.
19. Jack Regan, Julie Bryant, and Craig Weinschenk, *Analysis of the Coordination of Suppression and Ventilation in Single-Family Homes* (Columbia, MD: UL–Firefighter Safety Research Institute, 2020).
20. Gray, *FMFM-1: Warfighting*; and Krulak, *MCDP 1-3: Tactics*.
21. Anthony Avillo, "Uncommitted Tactical Reserve Assignment: The RIT Assist Team," *Fire Engineering* 173, no. 4 (April 1, 2020).

Chapter 6

1. Paul Grimwood, *Euro Firefighter 2: 6,701 Building Fires* (West Yorkshire, UK: D&M Heritage Press, 2017).
2. Grimwood, *Euro Firefighter 2*.
3. Grimwood, *Euro Firefighter 2*.
4. Fire Department of New York (FDNY), "Ventilation," *Firefighting Procedures* 1, no. 10 (2013); and Fire Department of New York (FDNY), "Ventilation," *Firefighting Procedures* 1, no. 10 (2019).

5. Dave Grossman, *On Combat: The Psychology and Physiology of Deadly Conflict in War and in Peace* (Warrior Science Publications, 2008); Robert Brown, *Smart 4 Life* (lecture, 36th Annual James Curran Memorial Seminar, Bureau of Training, Fire Department New York, New York City, April 16, 2016); Robert Brown, "Auditory Exclusion and How the Sympathetic Response Affects Our Operational Performance on the Fireground," *With New York Firefighters* 2 (2016).
6. Stephen Kerber, *Top 20 Tactical Considerations from Firefighter Research* (Columbia, MD: UL–Firefighter Safety Research Institute, 2016).
7. Stephen Kerber, *The Impact of Ventilation on Fire Behavior in Legacy and Contemporary Residential Construction* (Columbia, MD: UL–Firefighter Safety Research Institute, 2010).; and Stephen Kerber, *Study of the Effectiveness of Fire Service Vertical Ventilation and Suppression Tactics in Single Family Homes* (Columbia, MD: UL–Firefighter Safety Research Institute, 2013), https://ulfirefightersafety.org/docs/UL-FSRI-2010-DHS-Report_Comp.pdf.
8. Emanuel Fried, *Fireground Strategies* (Saddle Brook, NJ: Fire Engineering Books & Videos, 1972).
9. Craig Weinschenk and Robin Zevotek, *Exploratory Analysis of the Impact of Ventilation on Strip Mall Fires* (Columbia, MD: UL–Firefighter Safety Research Institute, 2020).
10. Kerber, *Top 20 Tactical Considerations from Firefighter Research*.
11. Keith Stakes, Julie Bryant, Nicholas Dow, Jack Regan, and Craig Weinschenk, *Analysis of the Coordination of Suppression and Ventilation in Multi-Family Dwellings* (Columbia, MD: UL–Firefighter Safety Research Institute, 2020), https://ulfirefightersafety.org/docs/Coord_Tactics_Multi_Family.pdf.
12. Andrew Fredericks, "Little Droplets of Water: 50 Years Later, Part 2," *Fire Engineering* (March 2000).
13. Josh Materi, "Protecting the Search," Fire Department Training Network, 20, no. 3 (2016): 12–14.
14. FDNY, "Ventilation" (2013); and FDNY, "Ventilation" (2019).
15. FDNY, "Ventilation" (2013); FDNY, "Ventilation" (2019); and Nicholas Papa, "Tactical Ventilation: Bridging the Gap between Research and Reality," *Fire Engineering* 170, no. 3 (March 1, 2017), https://www.fireengineering.com/2017/03/01/297459/tactical-ventilation-bridging-the-gap-between-research-and-reality/#gref.

Chapter 7

1. Firefighter Proving Ground, "FFPG Tips: Communication," Instagram (October 5, 2019).
2. Jack Regan, Julie Bryant, and Craig Weinschenk, *Analysis of the Coordination of Suppression and Ventilation in Single-Family Homes* (Columbia, MD: UL–Firefighter Safety Research Institute, 2020).

3. John Salka Jr., "Controlling the Door," *Firehouse* (October 31, 1996), https://www.firehouse.com/operations-training/article/10545569/controlling-the-door.
4. Stephen Kerber, *The Impact of Ventilation on Fire Behavior in Legacy and Contemporary Residential Construction* (Columbia, MD: UL–Firefighter Safety Research Institute, 2010); Stephen Kerber, *Study of the Effectiveness of Fire Service Vertical Ventilation and Suppression Tactics in Single Family Homes* (Columbia, MD: UL–Firefighter Safety Research Institute, 2013), https://ulfirefightersafety.org/docs/UL-FSRI-2010-DHS-Report_Comp.pdf; and Regan, Bryant, and Weinschenk, *Analysis of the Coordination of Suppression and Ventilation in Single-Family Homes*.
5. "Summary Data," Firefighter Rescue Survey (2019), https://www.firefighterrescuesurvey.com/summary-data.html.
6. Keith Stakes, Julie Bryant, Nicholas Dow, Jack Regan, and Craig Weinschenk, *Analysis of the Coordination of Suppression and Ventilation in Multi-Family Dwellings* (Columbia, MD: UL–Firefighter Safety Research Institute, 2020), https://ulfirefightersafety.org/docs/Coord_Tactics_Multi_Family.pdf; and Regan, Bryant, and Weinschenk, *Analysis of the Coordination of Suppression and Ventilation in Single-Family Homes*.
7. John Ceriello, "Training Minutes: Smoke Curtain for Flow Path Control," *Fire Engineering* (November 16, 2017), video, https://www.fireengineering.com/2017/11/16/287731/training-minutes-smoke-curtain-ceriello-2017/.
8. Stakes, Bryant, Dow, Regan, and Weinschenk, *Analysis of the Coordination of Suppression and Ventilation in Multi-Family Dwelling*; Jay Bonnifield, *The Anatomy of a Push*, webinar, September 24, 2020; Scott Corrigan, *Beyond the Door*, webinar, December 9, 2020.
9. Stephen Kerber, *Top 20 Tactical Considerations from Firefighter Research* (Columbia, MD: UL–Firefighter Safety Research Institute, 2016).
10. Kerber, *Top 20 Tactical Considerations from Firefighter Research*.
11. Jerry Tracy, "Welcome to the Ivy League," Fire Department Training Network, 18, no. 11 (2014).
12. Michael Ciampo and John Mittendorf, "Ventilation," chapter 14 in *Fire Engineering's Handbook for Firefighter I and II*, Glenn Corbett, ed. (Tulsa, OK: Fire Engineering Books & Videos, 2009).
13. Bob Pressler, "Truck Work at First Floor Fires," Fire Department Training Network, 18, no. 3 (2014): 1–4.
14. Bob Pressler, "More Truck Work: Second Floors, Attics & Basements," Fire Department Training Network, 18, no. 4 (2014): 10–12.
15. Stakes, Bryant, Dow, Regan, and Weinschenk, *Analysis of the Coordination of Suppression and Ventilation in Multi-Family Dwellings*.
16. Gary Klein, *The Power of Intuition: How to Use Your Gut Feelings to Make Better Decisions at Work* (New York: Crown Publishing, 2003); and Gary Klein, *Sources of Power: How People Make Decisions* (Cambridge, MA: MIT Press, 1998).
17. Scott Kleinschmidt, "Initial Search and Interior Isolation at Dwelling Fires," *Fire Engineering* 167, no. 1 (January 24, 2014).
18. Regan, Bryant, and Weinschenk, *Analysis of the Coordination of Suppression and Ventilation in Single-Family Homes*.

19. Peter Van Dorpe, *The New Fire Fight* (Tulsa, OK: Fire Engineering Books & Videos, 2018), DVD.
20. Kerber, *Top 20 Tactical Considerations from Firefighter Research*; Stakes, Bryant, Dow, Regan, and Weinschenk, *Analysis of the Coordination of Suppression and Ventilation in Multi-Family Dwellings*; and Regan, Bryant, and Weinschenk, *Analysis of the Coordination of Suppression and Ventilation in Single-Family Homes*.
21. Pressler, "Truck Work at First Floor Fires."
22. Mickey Conboy, "Practical Search Operations," (lecture, 36th Annual James Curran Memorial Seminar, Bureau of Training, Fire Department New York, New York City, November 5, 2016).
23. Sam Hittle, "Humpday Hangout: Survivability, Search, and VEIS," *Fire Engineering* (October 10, 2018).
24. Frank Ricci and Josh Miller, "Aggressive and Practical Search: It's Still about the Victim," *Fire Engineering* (March 1, 2016).
25. Michael Dugan, "Making a Window Entry for Ventilation or Search," *Fire Rescue* 4, no. 4 (March 31, 2009), https://firerescuemagazine.firefighternation.com/2009/03/31/making-a-window-entry-for-ventilation-or-search/.
26. John Salka and Rick Lasky, "Old School VES with Chiefs Rich Lasky and John Salka," Podbean (November 27, 2019), podcast, https://chieflasky.podbean.com/e/old-school-ves-with-chiefs-rick-lasky-john-salka/.
27. Bob Pressler, "The Firemanship Days," (lecture 2017), https://www.facebook.com/BrothersInBattle.
28. Brian Olson, "Vent, Enter, Search: Turning a Blind I," *Brothers in Battle* (December 1, 2015), https://www.brothersinbattlellc.com/blog/vent-enter-search-turning-a-blind-i.
29. "The Dangers and Deaths of VES," Search Culture, Facebook (March 28, 2018), https://www.facebook.com/SearchCulture.
30. Search Culture (January 26, 2021) 2020 Year In Review. Facebook.
31. Tim Klett, "Attic Fires," Fire Department Training Network, 20, no. 7 (2016): 1–4.
32. Daniel Troxell, "Primary Roof Ventilation Operations for Flat-Roof Structures," *Fire Engineering* 163, no. 8 (August 1, 2010).
33. John Jonas, "2086 Valentine Avenue, Bronx," *FDNY Pro*, 2 (2017): 8–11.
34. Matthew Murtagh, "Get the Roof," *Fire Engineering* (March 1991): 16.
35. Fire Department of New York (FDNY), "Ventilation," *Firefighting Procedures* 1, no. 10 (2013); and Fire Department of New York (FDNY), "Ventilation," *Firefighting Procedures* 1, no. 10 (2019); and John Norman, *The Fire Officer's Handbook of Tactics*, 4th ed. (Tulsa, OK: Fire Engineering Books & Videos, 2012).
36. Jerry Tracy, "Welcome to the Ivy League," Fire Department Training Network, 18, no. 11 (2014); Murtagh, "Get the Roof"; Bob Pressler, "Operations on Peaked-Roof Structures," *Fire Engineering* (September 1, 1995); John Vigiano, "The Ladder Company of the '90s," *FDNY Pro*, 2 (1999): 34–35; and Michael Scotto and Brian Mulry, "Top Floor Fires," *FDNY Pro*, 5, no. 56 (2020).
37. Ciampo and Mittendorf, "Ventilation."
38. Scotto and Mulry, "Top Floor Fires."
39. Norman, *The Fire Officer's Handbook of Tactics*.

40. Jonas, "2086 Valentine Avenue, Bronx."
41. Norman, *The Fire Officer's Handbook of Tactics*; Jonas, "2086 Valentine Avenue, Bronx"; Scotto and Mulry, "Top Floor Fires"; and Matthew Murtagh, "The Trench Cut," *FDNY Pro*, 2 (1981): 10–12.
42. Norman, *The Fire Officer's Handbook of Tactics*; and Ciampo and Mittendorf, "Ventilation."
43. Norman, *The Fire Officer's Handbook of Tactics*; and Ciampo and Mittendorf, "Ventilation"; Scotto and Mulry, "Top Floor Fires"; and Murtagh, "The Trench Cut."
44. Scotto and Mulry, "Top Floor Fires."
45. Nicholas Papa, "Overhaul Is Ventilation: Providing Air to Concealed Spaces," *Fire Engineering* (2019); digital version (May 25, 2020), https://www.fireengineering.com/2020/05/25/207305/overhaul-is-ventilation-providing-air-to-concealed-spaces-nicholas-papa/#gref.
46. Murtagh, "Get the Roof."
47. Tom Brennan, *Random Thoughts* (Tulsa, OK: Fire Engineering Books & Videos, 2007).
48. Conboy, "Practical Search Operations."
49. Mike Ciampo, "Training Minutes: Door Control, Size-Up and Forcible Entry," *Fire Engineering* (November 14, 2019), https://www.fireengineering.com/2019/11/14/480882/training-minutes-door-control-and-size-up-and-forcible-entry/#gref.

Chapter 8

1. Fred LaFemina, "Ventilation Basics," *FireRescue 1* (March 20, 2006), https://www.firerescue1.com/fire-products/ventilation/articles/ventilation-basics-rVoFqASavaSwz4gU/.
2. Aaron Fields, "The Nozzle Forward," (lecture, May 11, 2016).
3. D. H. Berger, *MCDP-7: Learning* (Washington, DC: US Marine Corps, Department of the Navy, 2020), https://www.marines.mil/Portals/1/Publications/MCDP%207.pdf?ver=2020-03-03-111011-120.

ADDITIONAL READING

Avillo, Anthony. *Fireground Strategies*. 3rd ed. Tulsa, OK: Fire Engineering Books & Videos, 2015.

Dugan, Michael. "Ventilation of Today's Fire Buildings is Crucial." *Firehouse* (May 21, 2007). https://www.firehouse.com/operations-training/article/10499468/ventilation-of-todays-fire-buildings-is-crucial.

Dunn, Vincent. *Collapse of a Burning Building: A Guide to Fireground Safety*. Saddle Brook, NJ: Fire Engineering Books & Videos, 1991.

Dunn, Vincent. *Command and Control of Fire and Emergencies*. Tulsa, OK: Fire Engineering Books & Videos, 1999.

Dunn, Vincent. *The Firefighter's Battlespace*. New York: FDNY Foundation, 2018.

Fredericks, Andrew. "Thornton's Rule." *Fire Engineering* (November 1996).

Kerber, Steve. Lecture. Andy Fredericks Training Days, Alexandria, VA (September 21, 2016).

Lombardo, Mike. "From Chaos to Coordination." Fire Department Training Network 21, no. 11 (2017): 1–6.

Mattis, James. Personal correspondence (2004). https://www.businessinsider.com/viral-james-mattisemail-reading-marines-2013-5.

Mitchell Jr., Douglas, and Daniel Shaw. *25 to Survive: Reducing Residential Injury and LODD*. Tulsa, OK: Fire Engineering Books & Videos, 2013.

Rhodes, David. "Science for Dummies Like Me." Parts 1–4. *Fire Rescue* (2017).

Rickert, David. "Full-On Assault." *Fire Engineering* 1, no. 2 (July 1, 2010). https://www.fireengineering.com/2010/07/01/295293/full-on-assault/#gref.

Roden, Erich. "Ventilation-Limited Fires in Residential Buildings." *Fire Engineering* 166, no. 7 (July 1, 2013). https://www.fireengineering.com/2013/07/01/253727/ventilation-limited-fires-in-residential-buildings/#gref.

San Francisco Fire Department (SFFD). *Ventilation Manual* San Francisco: SFFD, 2008. https://www.subburndown.com/pdf/official/SFFD_Ventilation_Manual.pdf.

Terpak, Michael, and Steve Woodworth. "We Have Not Been Doing It Wrong!" Fire Department Training Network 18, no. 12 (2014): 1–4, 14.181.

Willink, Jocko. *Leadership Strategy and Tactics*. New York: St. Martin's Publishing Group, 2019.

INDEX

A

accountability, establishing 79, 80
action plans 54
adaptability 60
aerosolized liquids 14
agility 59
air
    ambient 11
    entrainment 26, 44, 85
    exchange 22, 25, 29, 31
    flow 12, 22, 25, 28
    flow management 22
    insufficient supply 13
    inverse flow reaction 23
alligatoring 64
Anchorage (AK) Fire Department 120
Art of Reading Smoke 16
ash/soot 14
Asken, Michael 3
assessment, macro-level 6
auditory exclusion 5
autoexposure 86, 90

B

backdraft 13, 19
    explosion 105
barrel of the shotgun 87
below level 44
bidirectional flow. *See* flows, bidirectional
Boyd, John (US Air Force Colonel) 53
Braidwood, James
        (Superintendent) 11, 22, 25
Brennan, Tom (Fire Chief) 21
buildings
    downwind side of 85
    survey 14
    type II noncombustible 27
    upwind side of 85, 87–88
buoyancy 14
burning odor 14
buzzwords 23
by-products 11, 14
    accumulation 11
    migration 11

C

calorimetry 11
CAN (conditions, actions, needs) 77
carbon 14
    carbonization 19
    monoxide 14
    particulates 19
charlie-side wall 114
charring 14
chemical chain reaction 9
Chicago Fire Academy ix
Chicago Fire Department x
Clark, William (Battalion Chief) 9
combustion 9, 10, 11, 12, 16, 18, 21, 22, 31
    flaming and 9, 12, 15, 29
    fuel and 9
    hypoxic 13
    incomplete 14, 16
    noncombustible structures and 2
    siding and 90
    void spaces and 49
command presence 4
common materials 11
communication 68–70
    auditory exclusion and 69
    intelligence gathering and 68
    tunnel vision and 69
company-level commanders 5
compartmentalization 2
    of spaces 45
compartments 12, 26
components of ventilation 21–27. *See also* ventilation
    air flow 22–23
    fire spread 22–23
    managing inlets and outlets 23–27
condensation buildup 14
conditions, actions, needs (CAN) 77
confinement
    ability 11
confinement (*continued*)
    of the fire 55
confirmed
    entrapment 36

• 147 •

148 Index

life hazard 97
victim 71
corner trick 99

D

Davies, Frederick 100
Davis, Jimmy (Captain) x
Davoodi, Hossein 28
dawgs 98
deep seated fire 19
deflagration 45, 105
DiBernardo, Joey (Lieutenant) 1, 2
Dodson, David (Battalion Chief) 16
door control. *See also* doors
    containing fires and 82
    dangers of open doors 81
    extenuating circumstances and 82–84
    importance of 80
    opening doors and 80–81
    preemptive scans and 81–82
doors. *See also* door control
    as access points 85
    entry 82
    jacking 82
drafts 22
    stops 106
duct vents 108

E

education (of firefighters). *See also* training
    combining the mental and physical 3
    introductory curriculum for 1
    live burns and 2
    shortcomings of 2
energy
    efficient environment 12
    potential 10
    release 12
engine company 25, 79, 80, 82
    officer 77–78
entrapment 62
exhaust
    capacity 27, 31, 40, 43
    efficiency 24, 44, 80
    output 44
    point 17, 85
    pulsating 13
    unidirectional 38
exothermic reaction 22
exposures 90–92
    exterior 90
    hazards of 90–92
    interior 90
    protection from 55
extension, lateral 109
extinguishment 21, 26, 28, 32
    self-extinguishment 13–14

F

fatal gases 15
fire
    apartments and 78, 85
    attack stairwell 41
    BAG-ing 54
    contents 28
    deep seated 19
    direction of travel 72
    exhaust path of 85
    firebreaks 110
    floor 78, 85
    growth phases of 13
    lateral spread of 28, 32
    letting it blow 98
    mushrooming 23
    plumes 22
    rate of change of 58
    rate of growth of 72
    rate of progression of 6
    reaction time to 21, 29
    reenergization of 22
    seat 16
    spread, potential of 18
    suppression 25, 29, 81
    tetrahedron 9–13
    triangle 9
    vertical extention of 64
fire departments
    Anchorage (AK) 120
    London Fire Brigade (LFB) 11
    New Britain (CT) 155
    Parkersburg (WV) 23
    Washington DC 5
    Waterbury (CT) 21
Firefighter Rescue Survey 100

firefighters. *See also* education (of firefighters); firefighting
    outside vent 78
    roof 78
firefighting
    cognitive aspects of 3
    psychomotor aspects of 3
Firefighting Procedures 68
fireground 1, 3, 5, 10, 12
    footage 4
    tempo 58–62
fire seat area
    preparation of 80
fire tetrahedron, components of 9–13
flame, fingers and flickers 16
flashover 13, 16, 18, 21, 26
flashpoint 16
flows
    bidirectional 23, 24, 26, 28, 44, 80
    path 22, 81
    unidirectional 24, 26, 28, 85, 87
Fried, Emanuel (Deputy Chief) 23
fuel 9, 10. *See also* synthetic fuels
    availability of 26
    limited fire 28
    load 12, 14, 21
    natural 10
    rich 14, 18, 32
    source 2
    unburned 14

G

gas
    exhaustion 11
    fatal 15
    migration 22
    pressure 25
gasoline siding 92
Gay-Lussac's Law 22
glass 14
grace period 29
Gray, A. M. (General) xi, 1
Grimwood, Paul (Crew Commander) 53, 67

H

hazards (of exposure) 90–92
heat
    conditions 18
    outputs 10, 21
heat release rate (HRR) 10, 12
horizontal ventilation 27, 32, 43–47. *See also* ventilation; vertical ventilation
    advantages of 45
    compartmentalization and 45
    complications of 46
    creating inlets and outlets and 44
    vent as you go and 95
    of windows 45
HUD windows 46
Huggett, Clayton 11
human factors 3–7
    freelancing 4
    intuition 58
    misguidedness 4–6
    overzealousness 4
    tactics of isolation 68
    unawareness 4
HVAC (heating, ventilation, and air conditioning) systems 22
hydraulic ventilation 47–49. *See also* ventilation
    advantages of 48–49
    effectivity of 47, 48–49
    how to 47–48
    improving the smoothbore nozzle and 48
hydrocarbon-based materials 16, 19

I

ideal opening 39
ignition
    point 29
    temperature 9
incident commander 5, 61, 66, 79, 80, 82
inlet 23, 25, 26, 31, 44
    pure 85
intake pathway 38, 45, 81, 95
intuition 58
isolation tactics 68
itchy trigger finger 4

J
job performance 58

K
KO Fire Curtain 106

L
Ladders 4: Operations at Private Dwellings 44
latch strap 80
lateral extension 109
Layman, Lloyd (Fire Chief) 23
letting it blow 98
lifting effect 31
London Fire Brigade (LFB) 11

M
materials
    common 11
    hydrocarbon-based 16, 19
    naturally derived 16
    organic 11
    synthetic 16, 21
McAniff, Edward 28
McRaven, William (Commander of the Special Operations Command) 58
mechanical ventilation 49–50. *See also* ventilation
    correct execution of 50
    how it works 49
    pressurizing spaces and 49–50
    smoke ejectors and 49
moth-to-flame effect 16
mushrooming 23

N
naturally derived materials 16
navigation of void spaces 111–113
    checking for fire extension in 113–114
    handlines and 111
    intuition and 113
    knee-walls and 112
    relieving pressure buildup for 113
Neary, Tom (Deputy Chief) 57
negative pressure 12, 22, 47, 49
neutral plane 23, 26
New Britain (CT) Fire Department 155
Newton's third law of motion 25
New York Fire Department (FDNY) 23
Norman, John (Deputy Assistant Chief) 77
nothing showing 13, 14

O
observation holes 110
occupancy
    commercial 107
    McMansions 26
    multiple dwellings 107
odor of burning 14
off-gassing 9
on-deck crew 100
O pattern 26
openings 37–38
    availability of 12
    creation of 21
    how to control 72
    ideal 39
    location of 22
    low-pressure 27
    management of 21, 25
    natural 44
    on-level 26
    orientation of 12
    size of 22
operating autonomously 6
operations. *See also* standard operating procedures
    demands of 55
    immaturity in 4
    lens 64
    search and rescue 36
    systems 4
    tempo of 6, 58, 93
    window of 54
organic materials 11
organizational structure 4
outlets 23, 24, 26, 44
overhaul 111
oxidation 9
oxygen 9, 11, 12, 16
    concentration of 9, 12, 13, 16
    consumption of 12, 21
    deficiency of 13
    demand 12
    heat relationship 11
    supply of 11, 12

P

pawls 98
peak release rate 28
plastics 10, 19
positive pressure 12, 49
positive-pressure ventilation (PPV) 49
potential energy 10
PPV (positive-pressure ventilation) 49
preparation of fire seat area 80
pressure
    front 26, 44
    gradient 22
    negative 12
    positive 12, 49
pyrolysis 9, 16

R

RADE loop 53–57
    anticipate the progression 54
    determine the needs 55
    execute the tactics 56
    read the conditions 53
radiation feedback 18
rafters 43
rain roofs 108
rapid fire
    development 13
    growth 21
    progression 18
    spread 21
rapid intervention team (RIT) 62, 100
reflex time 58
returns 107
rollovers 16
roofs
    asphalt shingles on 92
    attics/cocklofts and 37, 38, 39, 40
    attic space and 44
    built-up on 108
    cockloft explosions and 11, 37–40, 43, 112
    fortified 108
    rain 108
    scuttles 108
    shaftway doors 108
    skylights 108
rotary saws 116

S

Scheidt, Richard xii
search and rescue operation 36
second fire growth stage 22
self-contained breathing apparatus (SCBA) 25, 79
self-extinguishment 13–14
self-vented 86
semantical confusion 23
situational awareness 5, 14, 57
slip-and-fall 41
smoke 11
    atypical presentation of 19
    cold smoke effect 19
    color of 18–19
    contamination 84
    curtains 84, 85
    density of 18
    discharge 16, 17
    ejectors 49
    exhaustion 11, 14, 26
    explosions 11
    heat-pressurized 17
    high-density, black 19
    ignition and 16
    laminar 18
    layering of 2, 26
    low-density, black 19
    odor of burning 14
    rate of change of 16
    suppression of 25
    temperature of 19
    thickness of 18
    toxicity of 14, 18
    tunneling of 28
    turbulent 18
    velocity of 17–18
    volume of 16–17
    volume-pressurized 17
smoldering 12
solid particulates 14
soot
    ash 14
    staining 14
spaces. *See also* void spaces
    survivable 55
    uncompartmentalized 82
spring clamp 80

stacking the deck 62
standard operating procedures 4. *See also* operations
  fireground playbook 5
subfreezing conditions 19
suppression 29, 81
survivable spaces 55
synthetic fuels 10, 11. *See also* fuels
  reaction speed and voracity of 11
synthetic materials 16, 21
synthetics, low-mass 15

T

tactical
  approach 55
  options 7
  ventilation 67–68
tenability 10, 18, 30, 31, 36, 41, 81, 84
Terpak, Mike (Deputy Chief) 67
thermal
  balance 23, 24, 25, 38, 82
  decomposition 9
  layering 2
  updraft 14, 25
Thornton's rule 11
Thornton, W. M. 11
three C's of communication 79–80
360-degree size-up 14
throat (of fire) 110
toxic
  dose 81
  production of gas 21
training 43. *See also* education (of firefighters)
  deficiencies 5
  examinations 2
traversing spaces safely
  ensure the integrity of pathways during 116
  exit plans for 117
trench cut 109
tripod position 116
tunnel vision 5
Turnell, Gregory (Lieutenant) 5

U

UL-FSRI coordinated fire attack study 29, 31, 81

underventilation 12, 13, 15, 16, 21, 25. *See also* ventilation
upper register 38
upwind side of the building 85, 87–88

V

vacant property security (VPS) systems 46
vantage point 14
vaporization 9
vapors 19
VEIS (vent/enter/isolate/search) 45
vent-enter-search (VES) 36, 72
ventilation 21. *See also* components of ventilation; horizontal ventilation; hydraulic ventilation; implementing ventilation; mechanical ventilation; underventilation; vertical ventilation
  communication and 77–79
  confinement ability and 11
  implementation of 27–33
  induced decay and 13
  induced flashover and 13
  limited fire 12
  local 45
  point 80
  positive-pressure ventilation (PPV) 49
  postextinguishment and 48
  preemptive 28
  premature 29
  profile 54
  selecting openings for 88–89
  shortcomings of 28
  survivability of victims and 30–33
  tactics 67–68
  timing of 28–29
  topside 38
  window of effectiveness for 11
  windows and 44
ventilators 108
venting
  for access 37
  for extinguishment 35, 78, 92–94
  for search 36, 78, 95–101
  self-vented 86
  tactical 119

vent point 78
    ignition 16
vents (duct) 108
vertical fire extension 64, 86, 107
vertical ventilation 3, 11, 28, 30, 43, 89
    complications of 39, 40, 41, 42
    creating the opening and 38
    detrimental factors for 38–39
    for extinguishment 103–110
    favorable conditions for 42
    purpose of 40
VES (vent-enter-search) 36, 72
victim survivability 1, 6, 10, 14, 23, 36, 57, 68
visibility 7, 38, 82
    zero 64
void spaces 2, 11, 37. *See also* navigating void spaces
volatility 21

W

wall cladding systems 92
Walsh, Charles (Deputy Chief) 35

Warrior Mindset 3
Washington DC Fire Department 5
water
    application of 11, 22, 32
    pressurized extinguishers 82
Waterbury (CT) Fire Department 21
WCD (wind-control device) 85
weeping windows 14
Willink, Jocko (Commander SEAL Team 3) 119
wind 13
    conditions 69, 78
    impact of 17
    velocity 22, 84, 89
    wind-impacted conditons 84–85
wind-control device (WCD) 85
windows
    horizontal ventilation of 45
    HUD 46
    ventilation and 44
    weeping 14
wing-type design 109
wooden structural elements 19

ABOUT THE AUTHOR

Nicholas Papa is a lieutenant with the New Britain (CT) Fire Department, where he has served for more than 13 years (6 of those as a company officer). He is presently assigned to Engine 1, New Britain's busiest fire-duty unit, previously working on Ladder 2 as a private. A second-generation firefighter, Nick entered the fire service in 2003, volunteering for a department in a neighboring suburb for 4 years.

Nick is a member of the UL-FSRI technical panel for the multiyear research project, "Study on Coordinated Fire Attack." He serves as a technical advisor and frequent contributor to *Fire Engineering*. In addition to authoring several articles, Nick has also produced the Training Minutes series, *Venting Tactically*, as well as the DVD *Lights Out: Knowing the Way When You Can't See*. Nick serves as cohost for the *Politics & Tactics* podcast. Nick has been an FDIC classroom instructor since 2017 and has taught at local and regional fire conferences across the country. He is also the sole proprietor of Fireside Training. Nick holds an MPA in Emergency Management from Anna Maria College and a BS in Public Safety Administration from Charter Oak State College.

Courtesy: Patrick Dooley